NATURAL GAS IN THE UK: OPTIONS TO 2000

NATURAL GAS IN THE UK: OPTIONS TO 2000

Jonathan P. Stern

Gower

Published by
Gower Publishing Company Limited
Gower House, Croft Road, Aldershot
Hants GU11 3HR, England

Gower Publishing Company
Old Post Road, Brookfield
Vermont 05036, USA

Reprinted 1987

British Library Cataloguing in Publication Data

Stern, Jonathan P.
    Natural gas in the U.K.: options to 2000.
    ----- (Energy papers; 18)
    1. Gas, Natural ----- Great Britain 2. Gas
industry ----- Great Britain
   I. Title II. Series
338.2'7285'0941   HD9581.G72

ISBN 0 566 05018 8

Printed and bound in Great Britain by
Halstan & Company Limited, Amersham, Bucks.

Contents

List of Tables

Foreword

The original intention when starting this study was to write a 2nd edition of Gas's Contribution to UK Self-Sufficiency, Energy Paper No.10. Although only two years have passed since that report was sent to the publishers, events of considerable significance have taken place, notably the Government's rejection of the import of Sleipner gas from Norway and the announcement of the intention to privatise the British Gas Corporation. The inquiries of the House of Commons Select Committee have enabled a multitude of opinions and a very considerable amount of hitherto unpublished data to be brought into the public domain.

The subject of energy self-sufficiency still commands public attention. However, the Joint Energy Programme has explored this subject very thoroughly in Papers 9-13 and in the book published in the Joint Studies in Public Policy series.* In the gas industry the focus of interest has shifted; other issues have become more urgent and, therefore, it seemed appropriate to widen the focus of the study.

The original paper was one of the series on UK energy self-sufficiency and was specifically written to fit with the other studies. No such constraint was present in the revision of the text. While the structure and some of the original text of Energy Paper No.10 remain, the demand projections and, most important, the self-sufficiency scenarios have been replaced by five scenarios reflecting different supply demand and price conditions.

We have had the benefit of advice from study groups of individuals expert in their particular fields and our thanks are due to those who took part. However, the views and interpretations presented are those of the author, and not necessarily those of the members either of the study groups or of the institutions that sponsor the Joint Energy Programme.

Robert Belgrave

*Robert Belgrave and Margaret Cornell (eds), Energy Self-Sufficiency for the UK?, Aldershot: Gower, 1985.

Summary

In the latter part of the 1980s, the UK natural gas industry faces a number of major decisions which will have an important bearing on its future. This study suggests that decisions about supplies, both domestic and imported, and the size of gas demand up to the end of the century, will be the critical issues. By contrast, the privatisation of the British Gas Corporation, which from the public perspective, appears to be of paramount importance, is unlikely in itself to have very far reaching consequences for the nation's gas supplies.

On the question of reserves, despite upward revision of estimates in the mid-1980s, there is no room for complacency. There have been no finds over the past decade to match those of the original Southern Basin gas fields which gave rise to the natural gas era in the UK. Smaller dry gas developments in the Southern Basin are encouraging, but will be developed at a higher unit cost. Condensate fields, which comprise one quarter of remaining proven and probable reserves, and one third of remaining recoverable reserves, present particular difficulties. Questionmarks surround the mode of development and financial incentives for starting condensate production, in a period of static or falling oil prices. Significant volumes of gas from these fields will need to be coming on stream, starting in the late 1990s, to compensate for declining production in the Southern Basin.

The availability of natural gas from the UKCS plus pipeline imports from a variety of sources means that liquefied natural gas imports will not be an attractive commercial option this century. Substitute natural gas produced from coal is unlikely to be able to compete with natural gas from any source until the second or third decade of the next century. Up to the end of the century, imports will probably be confined to Norway, with the possibility of small quantities from the Netherlands. There is increasing recognition that large-scale gas imports from the USSR could be a commercially attractive option in the 1990s and certainly in the next century, but uncertainty about the political acceptability of this source. Despite constant references by producers to the desirability of exporting UK gas, for the mid to late 1980s this is a dead issue. Even if producers were holding gas reserves which they genuinely wished to export to the Continent, it is very doubtful whether such gas could compete at attractive prices with Soviet and Dutch alternatives.

The most immediate decisions to be taken in the period up to 1990, given the lead times for production and import projects, relate to the mix of domestic and foreign gas supplies up to the end of the century and beyond. In this connection, the Government decision to reject the import of Norwegian Sleipner gas and exclude the possibility of imports up to 1995, was a landmark for the industry. If the present position on imports is maintained, most of the country's <u>proven</u> resource base will have been produced by 2000. At that time, an alternative would be to reduce the size of the gas market by a combination of: higher prices, a smaller industrial market and refusing to extend the reach of the grid.

	UKCS Production	Imports	Demand
	(billion cubic metres)		
1984 (actual)	40.2	13.6	53.8
1990	45	10	55
2000	45	10	55
2010	40	15	55
2020	35	20	55

The Table above is not a forecast. It presents what, to this author, seems to represent a reasonable compromise for all interested parties in the UK gas industry:

<u>For the consumer and the nation</u> - by maintaining the market at its present size; depleting the resource base at a brisk, but not irresponsible, rate over the next several decades, leaving a reserves to production ratio of more than 14 years in 2020; and ensuring that there is a gradual transition to greater import dependence from around 25% in 1990, to 35-40% in 2020.
<u>For producers</u> - by maintaining a high level of UKCS production throughout the first decade of the next century.
<u>For the utility</u> - by allowing the choice between UKCS production and imports, given a large and stable market size.

The scenario would provide the nation with the present level of gas supplies, while maintaining incentives for producers to continue high levels of exploration and production on the UKCS. Up to the end of the century, there is the option to meet demand by rapid domestic development of reserves and excluding imports. However, if imports are excluded and unless considerable quantities of additional dry gas reserves are located, which can be produced at costs similar to fields which are being brought on stream in the mid-1980s, the most likely outcome will be either a much larger dependence upon imports after 2000 or a much smaller gas market. Assertions that the country can be self-sufficient in gas up to the end of the century ignore the fact that in this sector of the energy balance it is essential to take a longer time horizon.

What is proposed here is not an all-embracing, inflexible strategy for the gas sector over the next several decades. It leaves room for changes of direction in order to accommodate changing availability of supplies (both domestic and imported), and changes in demand. The main hope must be that the option to deplete the UK resource base at the maximum possible rate will not be adopted simply because it may be the easiest course of action from a short- to medium-term financial point of view. The worst possible outcome for the gas sector would be very rapid rate of production over the next 15-20 years, followed by the development of major dependence on imports or a dramatic reduction in the size of the market. This would be particularly unfortunate given the opportunities available in the late 1980s to arrange a mix of indigenous production and imports from secure sources, which would ensure that the natural gas sector need not become an area of vulnerability in the nation's energy balance over the next three decades.

x

Glossary

Dry, non-associated natural gas: natural gas originating from structures where only gas is produced.*

Associated natural gas: natural gas originating from underground structures producing both liquid and gaseous hydrocarbons. The gas may be dissolved in crude oil (solution gas), or in contact with gas-saturated crude oil (gas cap gas). In such structures, gas production rates will depend on oil output, with oil usually representing the major part of energy equivalents.*

Natural gas liquids (NGLs): those hydrocarbons which can be extracted in liquid form from natural gas. Invariably the term NGL embraces propane and all heavier hydrocarbon fractions, ie butane, pentane, etc. In some instances it may be taken to include ethane as well*.

Gas condensate: the heavier NGLs - pentanes (C_5H_{12}) and hexane (C_6H_{14}), which would be liquid at normal temperature and atmospheric pressure, but have low boiling points which cause them to become vaporised and mix with other gases.**

***Proven reserves: those reserves which on the available evidence are virtually certain to be technically and economically producible.

***Probable reserves: those reserves which are estimated to have a better than 50 per cent chance of being technically and economically producible.

***Possible reserves: those reserves which at present are estimated to have a significant but less than 50 per cent chance of being technically and economically producible.

Load curve: a graph in which the send-out of a gas system is plotted against intervals of time.*

Base load supplies: supplies which are required throughout the entire operating period of a gas system.

Peak load supplies: supplies which are required only when demand reaches a daily or seasonal peak.

UKCS: United Kingdom Continental Shelf.

BGC: British Gas Corporation.

* These definitions are taken from Malcolm W. H. Peebles, Evolution of the Gas Industry. London: Macmillan, 1980, pp.211-15.

** Frank Frazer, Gas Prospects in Western Europe. London: Financial Times, 1981, p.9.

*** Department of Energy, Development of the Oil and Gas Resources of the United Kingdom (Brown Book), 1983.

Conversion Factors

One billion (thousand million) cubic metres (BCM) of natural gas per year is approximately equivalent to:

    0.04 trillion cubic feet per year
    100 million cubic feet per day
    375 million therms per year
    890,000 tons of oil per year
    17,800 barrels of oil per day

Where price conversions have been made between pence per therm and US $ per million British thermal units ($ per mmbtu): 10p per therm is approximately equivalent to $1.60 per mmbtu (where £1 = $1.60).

A Note on Privatisation

As this study is being completed, legislation to privatise BGC is passing through Parliament with the flotation expected to take place in the autumn of 1986. The House of Commons Energy Committee's report, <u>Regulation of the Gas Industry</u>, was published in early 1986, containing a wealth of material about the issues raised by privatisation and the form of the new regulatory bodies, the Office of the Director General of Gas Supplies and the Gas Users' Council, which will be created to oversee the industry.[1] At the end of February 1986, it seemed conceivable that changes in the political landscape could delay privatisation until after the next election. However, this note considers broad consequences of the legislation, assuming that it is passed in its present form. The details of individual issues, such as pricing, can be found in the text.

Thus far, criticism of the legislation has concentrated on: the form of privatisation which the Government has selected for the Corporation; the speed at which it is being implemented; and the motives behind both the form and the speed of the legislation.[2] The arguments in this area are concerned with the different options available to the Government in privatising the Corporation and particularly the options for splitting up the utility into independent regional and functional units. Since the public debate on these matters took place after the decision to privatise had been taken, it seems useful only to observe that the Government adopted the form of privatisation which would allow legislation to be passed as rapidly as possible. This has led to the feeling that a prime objective - and some have claimed, the major objective - of this privatisation has been the raising of revenues from the flotation within the shortest possible time.

This study is primarily concerned with privatisation to the extent that the flotation of the Corporation will bring about changes in gas supply, demand and trade policies of the utility. In this respect, it is too early to make any definite judgements. The Corporation is making the transition to the private sector in its present form, so that any major changes in its mode of operation may not become apparent for several years. However, at the beginning of 1986, it was difficult to see any substantial short-term changes in the gas sector which would be brought about by the flotation of the Corporation. Despite claims by the Government that greater competition would be introduced into an industry 'freed from Government intervention', there was general agreement with the conclusion of the Energy Committee, that: [3]

The Committee thus takes a far less optimistic view than the Secretary of State or the Chairman of BGC on the strength of competitive forces in the markets where gas competes. Far more detailed and convincing evidence than that so far offered would be needed to allay our scepticism. Consequently we believe that the regulatory regime must be genuinely effective both in the promotion of competition and in compensating for its weakness. If the regulatory regime proves ineffective, then a serious misallocation of national resources could result.

Following the publication of the report, a member of the Energy Committee successfully introduced an amendment to the Gas Bill which specifically gives the Director of Gas Supplies a duty to promote competition in the industry.

As far as the utility is concerned, the most far-reaching change which may take place relates to its view of the profitability of various activities. Having assessed the profitability of its constituent parts (functional, regional and sectoral), the Corporation's attitude towards future profitability in these areas will be important, particularly if its assessment conflicts with the present provision of services in certain functions, regions and sectors. This is likely to be most significant in the medium- to long-term, particularly if changes in availability of supply, or growth in demand, should lead to pressures either to expand or contract market size in the future.

Undoubtedly one of the major issues is the size and power of the regulatory authority, where some of the detail of the legislation has yet to be clarified. There is a division of views between those who broadly support the American model and those who believe that a small regulatory body should be the aim.[4] The critical issue is whether a small regulatory body, with limited power and resources, will be able to promote and safeguard the interests of the consumer to the necessary extent. Those who rightly criticise the American system because of its protracted and incredibly complicated bureaucratic procedures, tend to miss the point that, if the interests of consumers are to be fully taken into account, such arrangements may be inevitable.[5] In the British system, control will be exerted by means of regulation of price, rather than the American system of regulation of profitability. This basic difference should reduce at least a part of the bureaucratic procedures which make the American system so unattractive. However, there is concern that the powers of the regulatory body may be too weak; there are large areas in which it has no powers of oversight. Furthermore, a staff of 80-100 (including both the Office of Gas Supply and the Gas Users' Council) may be too small to carry out independent analysis in a large number of areas, as well as being greatly dependent for its judgements upon information provided by the utility and the producing companies.[6]

Some of these criticisms may be premature, given that we have not yet seen the full details of the legislation and it is hard to judge how it will work out in practice. The Gas Bill appears to give the Secretary of State wide powers of discretion either to act on his own initiative or to empower the Director General of Gas Supplies to take appropriate actions in a range of areas. The impression of this author is that, if the Office of Gas Supplies is to oversee a large number of areas other than pricing, it will need a much larger staff than is presently anticipated, with specialised regulatory skills which will take some time to develop. If the Office concerns itself mainly with the issue of price, which is probably inevitable in the early years of its existence, the utility will be very largely a self-regulating organisation which will only be challenged if there is public concern about its activities.

[1]   House of Commons Energy Committee Session 1985-86, _Regulation of the Gas Industry_. HC-15, 15 January 1986.

[2]   It is important to differentiate between those opposing the principle of privatisation and those opposing the form favoured by the Government. For examples of the former, see the evidence of the Trades Unions, in _Ibid_, pp. 120-141. For examples of the latter, the evidence of (among others) Colin Robinson and Eileen Marshall, and the Institute for Fiscal Studies, pp. 161-170.

[3]   _Ibid_, p. xiv.

[4]   See the evidence of Alex Henney, _Ibid_, pp.147-161

[5]   See the criticisms of American commentators in: Max Wilkinson, 'Regulation: vital but difficult to get right,' _Financial Times_, 29 November 1985; 'A Case for the US System,' _Ibid_, 5 December 1985.

[6]   This is the size of staff anticipated in the 1985 Gas Bill, p. vii.

Introduction

This paper takes a detailed look at the future options for the UK natural gas sector in the period up to the year 2000, and sets the scene for the first two decades of the next century. It reviews the various supply options, indigenous and imported, which may be possible in terms of resources and appropriate in terms of costs; the decisions which will need to be made on each of the supply options; and the timing of such decisions. Different demand options are also considered, based on different assumptions of the availability and cost of supplies and competition from other fuels.

The definitive history of the British gas industry concludes with the following statement:[1]

> From the outset, before one therm of North Sea gas was burned in Britain it was certain that the quantity available was limited. How long it would last depended on two main factors: firstly the total quantity ultimately available and secondly the rate at which it was extracted. Neither factor was then known with any precision, nor is it now: the only certainty was that there must be a time limit of some sort. Before that limit is reached - or rather, when its timing becomes apparent - some new strategy must have been developed to ensure continuity of supply.

In the mid-1980s, Government, producers and the British Gas Corporation have been involved in an intense debate about two major issues: the quantity of available reserves and the rate at which they should be extracted, combined with the option of gas imports. These issues have rarely engaged the attention of the general public which, since May 1985, has been preoccupied by the intention to privatise BGC within the life of the second Thatcher Administration.

The argument of this paper is that the issues of resources, production, demand and trade are of the utmost importance, not just for the gas sector but for the UK energy balance as a whole. There are a number of different strategies available for balancing gas supply and demand. This paper will outline some of those strategies, indicating the direction of current policy and the options which may be available in the future.

Natural Gas from the UK Continental Shelf: Resources and Production

Current UK natural gas production is concentrated offshore in the North Sea. In 1985, production also began from the Irish Sea where the Morecambe field is being used on a seasonal basis to meet peak winter demand. There are a number of areas which suggest considerable promise for the future, notably around the Shetland Islands where significant quantities of gas have been located in very deep water. Small natural gas finds have also been made onshore, and as costs of production offshore continue to rise, the attractions of land-based resources, even of small size, may become steadily greater.

This paper will deal almost exclusively with the resources of the UK North Sea. Inevitably there are differences in estimates of gas reserves on the UKCS and some of these are shown in Table 1. In each case, it is important to be clear on the definition of the figures, since different organisations use different classifications; even the UK Department of Energy 'Brown Book' classification changes slightly from year to year.[2] Given the uncertainties, the divergences in Table 1 are perhaps not as great as might be expected. In the initial proven and probable categories (or those categories which can be described as roughly equivalent to the Brown Book terminology), there is good agreement within a range of 1800-2200 BCM. Greater difficulty is encountered in trying to estimate possible or speculative reserves which may be economically recoverable, and it is likely that the sources are not describing the same phenomena. Here the figures, additional to the proven and probable reserves, are in the range of 500-800 BCM, with the only dissenter being Phillips Petroleum which gives a figure of 1500-2000 BCM in excess of its proven and probable estimates.

Looking at these round numbers in the context of the time frame of this paper, with the present level of production of some 40 BCM annually, remaining proven reserves would be sufficient to last until the year 2003 and, assuming probable reserves of another 600 BCM, the same level of production could be maintained for a further 15 years. Of course, these rather simple arithmetical calculations provide only the most rudimentary 'rule of thumb'. A popular conception of the (oil and) gas resource base is that of a large underwater gasometer out of which resources are drawn, with a gauge indicating how much gas remains to be produced. It comes as a disappointment to discover the complication of reserves calculated from a very large number of different fields - oil, gas and condensate - many of which may be at an early stage of delineation. Table 2 provides the Department of Energy breakdown of natural gas resources and emphasises the importance of distinguishing between the different groups of fields in different locations at different stages of development.

Dry gas

At the end of 1984, the Southern Basin of the North Sea, from which the vast majority of current UK gas production is drawn, contained remaining proven reserves of 425 BCM, while proven remaining reserves in the Frigg and Morecambe fields totalled 143 BCM. All of these 568 BCM of reserves are in fields which are already 'in production or under development'. At present rates of production the proven reserves of the Southern Basin fields would be exhausted in 18 years. The probable and possible reserves from the Southern Basin, as reported in the 1985 Brown Book, would extend production by an additional 18 years at current levels. Even these approximations are misleading since fields such as Morecambe will be used for peaking purposes only, thus curtailing further the period of uninterrupted base load supply from the area.

The Southern Basin fields listed in Table 3 carried the UK into the natural gas era. Five major accumulations have been producing for nearly a decade and all have a reserve life of around another decade at present rates of production. Realistically, this means that by the early 1990s production from these fields will be on a slowly, and perhaps a steeply, downward path. Of the other gas fields which have been located thus far on the UKCS, only Frigg is comparable in size with the larger Southern Basin accumulations, although the field is far less conveniently located (especially since 60% of the gas is under Norwegian jurisdiction). As the currently producing Southern Basin fields become exhausted over the next 10 years, replacement reserves, even those presently known in the Southern Basin which will extend the life of the province by 5-10 years at the present rate of production will be more difficult and expensive to develop.

Table 4 shows the reserves in individual Southern Basin fields which are in production and under development, while Table 5 contains estimates of probable and possible developments in the period up to the mid-1990s. The total of 741 BCM in fields in production and under development in Table 4 is almost identical to the Brown book figure of 739 BCM for proven and probable reserves (Table 2). The Wood Mackenzie estimates in Table 5 suggest a total of 282 BCM of reserves in Southern Basin fields which could be producing before 2000, which compares with the Brown Book's proven and probable figure (in other significant discoveries not yet fully appraised) of 365 BCM. It is therefore reasonable to assume that the fields listed in Tables 4 and 5 contain more than 90% of the Department of Energy's estimates of proven and probable reserves in the Southern Basin.

One feature of these statistics in particular stands out: the five large fields in the Southern Basin initially contained reserves ranging from 50 to 300 BCM. Of the fifteen Southern Basin fields (or groups of fields) under development, or with a probability or possibility of being developed in the period up to 1995, one (Valiant) has reserves of 55 BCM, one (Audrey) has reserves of 34 BCM, five are in the 20-30 BCM range, five are in the 10-19 BCM range, and three have single figure reserves. Making some assumptions about the annual production rates of the fields in Table 5, it appears that these fifteen fields would yield around 20 BCM per year, compared with the present production of 30-35 BCM per year.

Of the other dry gas fields, the UK section of Frigg field, with some 60 BCM of remaining recoverable reserves, will be exhausted by 1993.[3] The Morecambe field in the Irish Sea, containing some 140 BCM of proven reserves, has the capacity to deliver a minimum of 600 million cubic feet per day (mcf/d) - equivalent to a yearly delivery of 6 BCM - increasing to a maximum of 1,200 mcf/d by the end of the decade.[4] It is planned, however, to release these volumes only at the time of maximum demand during the winter months. However, one of the notable features of the Morecambe field is that its mode of use could be changed from peak to base load (and vice versa) to accommodate the changing supply and demand requirements of BGC. Thus it is difficult to put an estimate on the life of the Morecambe reserves, or to use these reserves when calculating the base load supply availability. Apart from 42 BCM of probable and possible reserves in the Frigg and Morecambe fields, the remainder of the reserve base is to be found either in condensate fields or in fields where the gas is associated with oil.

Associated gas

Remaining proven reserves of associated gas amount to 118 BCM, with 85 BCM of probable and possible reserves in equal quantities. Natural gas found in association with oil and natural gas from condensate fields gives rise to a number of problems: technical, logistical, commercial and fiscal. The most obvious logistical obstacle arises where deposits are situated offshore in geographically dispersed locations. What is required is a number of pipelines connecting oil and condensate fields and feeding into a main gas trunkline to shore. This is generally referred to as a 'gas gathering pipeline' system.

Where gas produced in association with oil cannot be reinjected into the deposit or collected and piped to shore, it must be flared if oil production is to continue. Flaring of associated gas is a problem for oil producers worldwide; up to 1980, OPEC countries had been flaring more gas than they utilised.[5] It is probably fair to say that, only following the 1973 world oil price rise and the consequent rise in natural gas prices, has much of the associated gas worldwide become economic to gather. Even now, many would argue that the associated gas marketed by a number of the large gas gathering systems of Gulf oil producers will never realise an acceptable commercial rate of return on such a large investment.

While it might be possible for countries with small populations and low (or zero) domestic demand for gas to contemplate substantial flaring, for the UK, with a very substantial domestic demand (partly sustained by large-scale imports), the flaring of more than marginal quantities of fuel must be a matter of serious concern. Volumes of gas flared from offshore oilfields rose to a peak of 6.6 BCM in 1979 (Table 6), of which one half came from the Brent field where, as a result of technical problems and newly imposed government flaring restrictions, the operator was forced to make a temporary cut in the rate of oil production.[6]

The likelihood of increased flaring, combined with the advent of sharply increased oil prices (making economic a number of fields hitherto considered marginal) and an anticipated shortage of energy and natural gas, gave rise to an appraisal by the British Government of the different gas gathering options in the UK North Sea.[7] The option which received most serious consideration was the joint BGC/Mobil 'Northern North Sea Gas Gathering System', designed to transport natural gas and natural gas liquids (NGLs) from both the UK and Norwegian sectors to the UK mainland. The BGC/Mobil study concluded that the system would yield approximately 11 BCM of UK gas per year in the late 1980s and early 1990s, augmented by roughly the same quantity of Norwegian gas. The total quantity of UK gas available for collection was estimated to be some 150 BCM, in addition to which some 42 million tonnes of NGLs would be made available (3.4 mt in a peak year), which might be augmented in the 1990s to a total recoverable reserve in excess of 93 mt.[8]

The reasons why the system was eventually abandoned are complex, and different actors laid the stress and/or the blame on a whole range of factors and third parties. Certainly a considerable setback was the loss of Norwegian Statfjord gas to the Continental countries, and there was a view that, in the absence of Norwegian gas, the system was not economically viable. However, even without this problem, a vicious circle of attitudes had developed. The financial community would not lend funds - the estimate of which had risen from £1.1 billion to £2.7 billion sterling - without a throughput contract in existence between the oil companies and BGC, or a guarantee of repayment from the oil companies. In turn, the companies would not commit funds without a commitment from BGC on the price of the gas. BGC would not give a commitment on price without a volume guarantee from the companies. The logjam created by these positions threw the burden of financing back on the Government.

The Government of the day had intended the gas gathering systems to be operated as a privately-owned public utility, with the oil companies as the major shareholders along with BGC. A significant commercial obstacle to the creation of such a utility was that it would have had no immediate profits against which to offset its development costs and indeed would not have been able to see any such profits for some years to come. More substantial State financing, which had become the only possible option for keeping the project alive, was unacceptable to the Government which took the view that, since the private sector had refused to finance the project, the companies should 'make their own arrangements for bringing the gas ashore'.[9]

In effect, this left the Shell/Esso Far North Liquids and Associated Gas Systems (FLAGS) as the sole gathering facility for UKCS gas. Originally designed to transport the associated gas from the Brent field to St Fergus in North-East Scotland, a 'western leg' including gas from the North and South Cormorant, Ninian and N.W. Hutton fields was added to the plans in 1978. A 'northern leg' was acquired in April 1981 when the gas from the Magnus, Murchison and Thistle oil fields was switched from the proposed gas gathering system (even before that scheme was abandoned).[10] Despite Norwegian Statfjord gas having been contracted to the Continental consortium, the UK Government and BGC refused a request from Statoil (the Norwegian State oil company) to allow the UK portion of the gas to be delivered to the Continent via the Norwegian Statpipe system, and insisted that the resource should be available to UK consumers. The FLAGS line was commissioned in 1982 and was raised to full capacity during 1985, including the first deliveries from Statfjord in October of that year.[11]

These developments greatly reduced the flaring of associated gas. Comparing 1979, the peak year, with 1984, volumes flared had fallen from 6.6 BCM to 3.4 BCM in a period when oil production had risen by more than 60%. Thus the fall is more dramatic when expressed in million cubic metres of gas flared per million tons of oil (mcm/mt) produced, falling from 83 mcm/mt in 1979 to 28.2 mcm/mt in 1984. This has been achieved very largely by the FLAGS line, plus joining other Northern North Sea fields (Piper and Tartan) to the Frigg line. In 1986, a line carrying gas and NGLs from

Fulmar, Clyde and eventually a number of other smaller fields will become operational. Initial volumes of gas will be small, probably less than 1 BCM per year.[12]

In 1984, the Brown Book listed eight fields where all gas not used on the platform is flared (a total of 59 million cubic metres per day) and two oil fields - Beryl A and South Brae - where reinjection is taking place and no current plans exist for transportation of gas to shore.[13] The remaining uncertainty is whether there is sufficient capacity within the FLAGS line to cope with the volumes available from oil fields currently in production and likely to be available in the short to medium term.[14] Table 5 contains three oil fields where associated gas production is expected by the early 1990s.

The collapse of the BGC/Mobil system raised a number of interesting questions about the organisation and financing of gas gathering efforts in the 1980s. In early 1984, there was another reminder of the difficulty of getting such schemes off the drawing board when a medium sized system involving 20 oil companies was shelved.[15] There is a strong argument that any size of gathering system needs a 'lead field' - a major gas source which will provide the majority of the throughput and the focus for others to be drawn in: Brent in the case of FLAGS and (Norwegian) Statfjord in the case of BGC/Mobil. Whether or not this is so, it seems clear that in the absence of a buyer who is prepared to sign up a number of sources in advance, with some assurance on prices, private funding will be impossible to arrange. This leaves the option of direct State financing, which seems unlikely to appeal to a government which has previously rejected such an arrangement.

In trying to plan a gas gathering system, there is the problem of ensuring that the individual field operators can co-ordinate development in order to dovetail with throughput availability in the system. There must be a question as to whether systems which involve a large number of operators are workable, even with maximum co-operation and goodwill on the part of those involved. The sheer complexity of negotiations between a large number of partners means that the lead times for reaching agreement to construct the system are likely to be extremely lengthy - probably longer than the actual physical construction of the system itself. In addition, a system which is designed to include a large number of fields probably has to take an optimistic view of (oil and) gas prices and hence the economic viability of fields which appear marginal at the planning stage.

In the 1980s, therefore, the organisation of gas gathering emphasises private rather than State-owned schemes, and on small-scale rather than large-scale systems. In practice, this is likely to mean that one operator has to take the initiative to build a line from a particular field to shore with some built-in spare capacity to allow for others to feed in at a later stage, rather than a large-scale private or State-owned system. Small-scale systems will only encompass those fields which look certain to yield an acceptable commercial return on the investment within a short period of time. Operators are more likely to want to put gas from their own fields through their own pipeline, rather than through that of a competitor. Spare capacity may be too expensive to build into a pipeline unless a future partner is willing to put up some investment ahead of time, or the fiscal regime makes some allowances for this.

Thus in the mid to late 1980s, less gas will be gathered in the short term by small schemes than would have been the case under the much larger

BGC/Mobil system.[16] This does not appear to be a problem at present and nothing is lost if the gas is gathered at a later stage. However, there is the danger that, under pressure to maintain and/or increase oil production, a proportion of associated gas will be flared which might otherwise have been collected and used. The situation will depend on the strength of the government commitment to a tight gas flaring policy. In turn, this commitment will be affected by the degree of government reliance on a high level of crude oil production and exports. The greater the dependence on high oil revenues, the greater the temptation to permit a high level of gas flaring.[17]

Condensate fields

Proven reserves in gas condensate fields amount to only 40 BCM, but another 283 BCM are probable and the possible resource base is larger still at 337 BCM. After the dry gas fields which have been described above, some of the largest UKCS accumulations of gas are to be found in condensate fields, some of which are shown in Table 5. There is considerable difficulty in discussing condensate fields under a single heading, since the value of the liquid fractions and the ratio of natural gas to liquids varies in each case.

In addition to the gas gathering and transportation issues, condensate fields raise other complex problems. There is a fundamental choice to be made in producing a condensate field: the gas can be produced at an early stage of development by 'blowing down' the liquid fractions (which means that a large percentage of the liquids will be lost), or it can be recycled in order to provide pressure for continuing liquids extraction. The recycling option delays gas production for a period of years, in most cases the best part of a decade. From a national point of view, there is an obvious attraction in producing the NGLs to the maximum possible extent. However, from a commercial point of view, there is uncertainty about the planning and profitability of the gaseous component, particularly in small condensate fields. Selecting a development option for condensate fields is therefore extremely complex and will depend on the composition of the hydrocarbons and the location of the specific field. If it is possible to make a generalisation, the inclination (which would require the agreement of the Department of Energy) would be to 'blow-down' small accumulations and recycle larger ones.

From the point of view of the individual operator, recycling condensate fields does not represent a particularly attractive commercial proposition. Illustrative figures in Table 7 suggest a recycling period of 8 years after the first oil production before the start of gas production. When the time comes to market the gas, the purchaser is well aware that investments have been made and that the seller is in a weak position in any negotiation. In this regard, the development of the first UKCS condensate fields, Brae and North Alwyn, where gas is not expected to be available until the 1990s, will be an important indication to aspiring operators, particularly for a field as large as Bruce (Table 5). At present, purely from the point of view of revenue planning and cash flow, the recycling option is unattractive and there may be a need to adjust the fiscal regime for a limited number of large condensate fields in order to encourage maximum production of liquids. This would suggest immediate gas production from a large number of smaller accumulations with consequent loss of the liquid fractions.

Nearly half of the remaining recoverable UK gas reserves is in oil and/or condensate fields. Of these non-dry gas resources, more than three quarters are in condensate fields, of which only 6% are in the proven category. The problems of creating access to associated gas, and particularly to condensate reserves, do not permit simple answers. The reason for discussing this part of the resource base in some detail is that these are the critical and uncertain elements in gas production in the mid to late 1990s, the resolution of which will have important repercussions on the availability and timing of UKCS gas production. Gas which is flared is lost permanently from reserves, while gas which has been located in a deposit (particularly a condensate deposit which will require many years of recycling), far from existing pipelines, may be some years away from being available for consumption. Indeed, the UK may have to face the possibility that development of gas in some fields will not be considered economic for a considerable period of time, perhaps several decades.

By 1986, the average annual capacity of associated natural gas pipelines to the UK mainland had reached some 8 BCM (6 BCM via FLAGS, up to an additional 1 BCM through the Frigg line, and 1 BCM through the Fulmar line).[18] In the early 1990s spare capacity will open up in the Frigg line, which may solve some problems for fields in convenient geographical locations. Pipelines bringing in imported gas would present additional opportunities. One of the attractions of the proposed import of Norwegian Sleipner gas was the opportunity to pick up gas from a number of oil and condensate fields along the route of the line, thus acting as another gas gathering pipeline.

No gas is currently being produced from condensate fields, yet these fields account for one quarter of proven and probable reserves and one third of the total gas resource base. Moreover, there are important technical, commercial and taxation issues which have still to be resolved before large-scale gas production from condensate fields is likely. Optimistic projections of UKCS gas production suggest that this will start in the mid 1990s.

Field development costs

The lack of a cost curve for current and future production from gas and condensate fields is an enormous handicap in assessing potential production rates and the extent to which resources which have been identified can be considered economically viable. Part of the difficulty with data stems from the need of both the oil companies and BGC to maintain a negotiating stance on appropriate prices (a subject considered below). However, although there is much genuine difficulty in assessing the extent to which costs have risen and are likely to rise in the future, there is no question that the costs of developing recently discovered dry gas fields, let alone gas from oil and condensate deposits, are of a different order of magnitude from the original Southern Basin gas discoveries.

Some illustrative capital and operating cost figures are shown in Tables 7 and 8. However, the House of Commons Energy Committee reported that:[19]

> The costs of developing the West Sole and Hewett Fields...which came on stream in 1967 and 1968 respectively, were between 4p and 5p per therm at today's prices and before tax and royalty. The Southern Basin fields that are now under development or being planned, on the other hand, are likely to cost between 15p and 20p per therm, again before tax and royalty. As smaller, more difficult and more distant reservoirs are

tapped, the costs will undoubtedly be higher still. BP and Phillips, for example, assume that the marginal cost of new gas developments at the turn of the century is likely to be around 30p per therm and could be as high as 50p per therm. This relentless increase in costs will need to be reflected both in the prices paid for gas at the beach head and in the fiscal regime. If it is not, then the pace of UKCS development will slacken and the propensity to import gas will be increased.

BP gave their most recent reserve estimates with direct reference to prices (as opposed to costs), claiming that 'at prices up to 27-31p per therm, the most likely discoveries and produceability would be 77 trillion (cubic feet).'[20] From these figures taxes and royalty have to be subtracted in order to gain an impression of average costs. There is the related issue of threshold sizes of fields, below which it is not economic to produce the gas. This is a complicated subject where the outcome is greatly dependent on whether the field is a 'stand alone' project or part of a satellite development, and whether it can use a third party pipeline or requires a new pipeline and shore terminal. Mid-1985 estimates by Morton Frisch were that, at a price of 25 pence per therm, Southern Basin threshold sizes would be 14-21 BCM for fields requiring new pipelines and terminals and 7-14 BCM for those which could use existing facilities.[21]

In addition, there must be concern in the mid-1980s about whether these developments can commence under conditions of falling (real and nominal) oil prices. Although such fields will not be entering production until the mid-1990s, when a completely different oil and energy price situation may obtain, this is likely to be a difficult climate in which to attract investments for such projects.

UKCS gas resources: summary

In the mid-1980s, a number of extremely bullish forecasts of UK gas reserves have been published, some of which were not unrelated to the views of their authors on the need for gas imports. Nevertheless, the mood of the industry at the present time is optimistic.

When 'new' (higher) gas reserve estimates for the UKCS are announced, it is always worth asking what has changed in order to give rise to these higher figures:

a)   new discoveries which have led to an increase in producible reserves in place;

b)   a more favourable tax regime which encourages companies to develop reserves identified some time previously;

c)   higher prices offered by buyers, which encourage companies to develop reserves previously identified.

In b) and c), the establishment of these conditions should be distinguished from a more favourable tax regime and/or a promise of higher gas prices, which may encourage greater exploration for reserves which are believed to exist, about which little is known.

The future economic viability of dry and associated gas resources depends on the cost of production, the price offered for the resource, and the government tax policy. The quantity of gas available at any given date will be a function of how operators see the levels and interplay of those three factors currently and also five to ten years hence. The price which is being offered for new UKCS gas will reflect the views of BGC (and other potential purchasers) of market conditions and the competitive position of gas vis a vis other fuels. Prices offered to purchasers, and hence the decision of the producer to go ahead with development, will therefore be importantly affected by considerations such as oil price movements. Thus the fall in oil prices during the latter part of 1985 and the beginning of 1986 will tend to slow down new UKCS gas developments, particularly if this is perceived to be the beginning of a period of lower oil prices lasting for several years.

Once it has been established that new reserves - as opposed to the promise of new reserves - have been discovered or created by an improved fiscal climate, there is then the question of lead times. For example, recent small finds in the Southern Basin could make a significant impact within a five-year time frame; for new finds in the Central Basin, the time frame would be closer to ten years. The first surveys were made for the FLAGS line in 1975 and became fully operational only in 1985.[22] Such factors must constantly be kept in mind when discussing reserve levels and plans for future production being made by the private sector. Uncertainty surrounding the size and cost of future domestic gas supplies may, paradoxically, be greater than for certain external sources of supply.

From a purely resource perspective there appears to be no problem in maintaining production up to 1991 at current levels. During the 1990s, the Southern Basin fields currently in production will start to decline. The Frigg field will start to decline in 1991 and once this process begins, production will fall to zero within two to three years, although there will continue to be a small quantity of gas produced from the Frigg satellite fields.[23] Around 1993, therefore, a gap will begin to open up between UKCS production and projected demand levels up to the end of the century.

There is support for the thesis that, on purely reserve considerations, this gap can be filled entirely from UKCS resources. However, this prospect raises two further questions: whether such a course of action is physically possible and, if so, whether it is strategically desirable, when considered in a time frame beyond the end of the century. On the question of physical possibility there is certainly a challenge of human and engineering resources. The UKOOA study of potential gas production up to the end of the century raises the overall ability of companies and supporting industries to implement viable development opportunities, as a potential problem.[24] However, more serious doubts relate to the lack of discoveries comparable in size to the original Southern Basin fields, to replace both those fields and Frigg as they begin to run down during the 1990s. In order to offset these developments a large number of small dry gas, associated gas and condensate fields will need to be developed, which will require long lead times and a more complex transmission system. This undoubtedly suggests a much higher cost of production and transportation for gas from future deposits; just how much higher may be the critical question for future production levels from the UKCS.

The second question of whether rapid production of UKCS resources is strategically desirable, can only be addressed with reference to the availability of external sources of natural gas.

Natural Gas Import Options

Pipeline options

Norway. Gas imports from Norway began in 1977 through a pipeline from the Norwegian sector of the Frigg field to St Fergus. The UK had earlier bid for the gas from the Norwegian Ekofisk field, but had lost out to the Continental European consortium of utilities from West Germany, France, Belgium and the Netherlands. The disappointment of losing Ekofisk gas was compensated by the contract for more than 149 BCM of Frigg gas. With some 68 BCM already delivered in the 1978-84 period, the contract allows for around 9-11 BCM of gas to be delivered annually up to 1990, with volumes running down to zero by 1993.[25] If the redetermination of the reserve base, carried out during 1985, proves to have been unfavourable, it is possible that production might turn down as early as 1988.[26]

In 1982, BGC began negotiations with Statoil for supplies from the Sleipner field which, in terms of volumes and commencement of deliveries, were intended to be a direct replacement for Frigg. The details of the Sleipner contract and the ensuing negotiations can be found elsewhere.[27] Briefly, in February 1984, after twenty months of negotiations, BGC agreed terms with Statoil. This contract had to be formally approved by the British Government but, despite the version of events which was circulated after the decision, this was never expected to be a problem.[28] After studying the draft contract, the Department of Energy (DOE) supported the deal with some minor adjustment to the delivery profile, and entered into direct negotiations with its Norwegian counterpart on the question of: the Sleipner NGLs, a specified level of participation from British contractors, and matters concerning the treaty. On 11 February 1985, however, Energy Secretary Peter Walker announced to Parliament that:

> ...proven and probable reserves...as shown in the Brown Book published in April 1984, have increased by 6.2 TCF. As a consequence, the Government has concluded that it will no longer be necessary to import gas in the 1990s on the scale anticipated even last summer. Accordingly, the Government has decided not to endorse ..the purchase of gas from the Sleipner field.

The Sleipner episode highlighted a number of issues in UK gas supply and trade policy to which we shall return later. However, it has contributed to a radical change in Norwegian gas export policy and a short digression is required at this point, because this policy will be of paramount importance to the UK for the foreseeable future.

Until late 1983, Norwegian gas export policy could be crudely characterised in the following terms: the country has a small population and a very large hydrocarbon resource endowment. Although the majority of that endowment is gaseous, there is also a comparatively large oil component. The problem with the gas resources is that, although they are huge, they are also high cost, i.e. the Troll field (to say nothing of the Northern Norwegian resource base) presents difficult geological conditions, in deep water, far from the mainland. Therefore, the argument ran, if importing countries wished to import Norwegian gas in large quantities, they would be required to pay considerably higher prices than those then obtaining on European markets. If gas importers did not wish to pay those prices, Norway would leave the gas option to a future date and switch to its 'oil option', in order to provide the country with continuing employment and industrial activity, and the revenues which it would need in the 1990s.[29]

During the final months of 1983, the Sleipner negotiations greatly reduced the price that Statoil had been hoping to get from BGC. This represented a retreat by the Norwegians but, from their point of view, much worse was to come when the British Government turned down the contract. The failure to conclude the contract represented a major setback for Norwegian gas export policy. This was only one factor which made the climate of gas exports look very much more unfavourable for the 1990s. Others were: the dramatic weakening of gas and energy demand and prices on Continental European markets (as well as a considerable reduction in demand forecasts for the period up to 2000); the renewal of all of the Dutch export contracts for a period of 10 years which meant that those deliveries would continue through the period 2005-2010; aggressive marketing by the USSR with its very large exportable gas surplus available at attractive prices. A combination of these factors gave rise to a major change in Norwegian attitudes towards the marketing of gas exports. This greatly affected the urgency in the marketing of gas from the Troll field. The loss of the Sleipner contract left the Norwegians with very little to look forward to in the 1990s as far as gas exports were concerned: Frigg and Ekofisk expiring in the early years of the decade, and only the comparatively small Statfjord exports of 3-6 BCM per year continuing after 1995. By mid-1985, Statoil had announced its intention to market the first phase of the Troll field, which would amount to some 15 BCM of gas per year starting in the mid to late 1990s.

By the beginning of 1986 the rationale behind Norwegian gas export policy had changed completely. Troll gas was no longer considered to be either high cost or frontier technology. It was simply a version of what already existed elsewhere in the North Sea, scaled up three times. Advances in offshore technology had shortened the lead time for the field from about a decade to around seven years. Moreover, Statoil appeared to have become convinced of the need to compete aggressively in West European gas markets in order to secure market share in the 1990s.[30] The Norwegians seemed to have accepted the position of the Continental importing consortium, that if they did not offer the gas at 'market-related prices', there would be no buyers.[31] While much remained to be settled regarding the Norwegian Government tax take, the determination to sell Troll gas at market-related, rather than cost-related, prices appeared to have been accepted. The phrase 'oil option' had disappeared from the Norwegian vocabulary.

So far as the UK is concerned, this is a happy turn of events. It appears that during 1986 a contract will be signed with the Continental consortium regarding Troll (Phase 1). It is believed that such a contract will be in the region of 10-15 BCM per year, with an option (or options) to increase volumes. While the UK has played no part in the Troll negotiations thus far, there must be every likelihood that, in the early 1990s, it could step in with an offer for a share of Troll gas. This is particularly likely, given the transportation logistics: it will be necessary for Troll gas to be marketed in Continental Europe through an expansion of the existing pipeline system. However, from 1993, the Frigg line will have a large amount of unused capacity and construction of an additional connecting line from the Troll field would not present a major problem. It may therefore be possible for a portion of Troll gas - perhaps as much as 10 BCM - to be delivered largely through the existing pipeline infrastructure. Alternatively, BGC might decide to revert to the original Sleipner deal assuming that this gas is still on offer, through a new pipeline as previously intended.[32]

The Norwegian resource base is such that the country will have the capability to sign contracts for additional base load supplies covering the next several decades. Supplies from Norway will arrive by pipeline, although the route of such lines is uncertain and, as indicated above, may be determined by the development of gas fields on the UKCS. It may be possible (and certainly less costly) to use existing lines, i.e. Frigg (gas) and Ekofisk (oil), as those facilities become idle with the exhaustion of resources. Another interesting prospect is the creation of a new pipeline system through the UK to be used as a transit route for Norwegian gas to Continental Europe; this will be discussed below.

Supplies via Continental Europe. The prospect of pipeline gas from the Netherlands was raised in the early 1960s, after the discovery of the Groningen field and prior to the finds on the UKCS.[33] Following the discovery of the Southern Basin fields, these ideas were abandoned and exports of Dutch gas were concentrated in the Continental market. With Dutch gas largely committed to export markets and the country already importing gas from Norway, the prospect of exports to the UK would now seem unlikely, were it not for the change in Dutch export policy in the early 1980s, consequent upon a reappraisal of reserves. The Netherlands discovered new reserves and uprated existing fields, which enabled Gasunie to extend large-scale export contracts beyond the end of the century when all of them had been due to expire.[34]

In addition, the Netherlands is the only European country which has the ability to act as a 'surge supplier' (or supplier of last resort) of major proportions, for Continental European countries which may become exposed to supply interruptions during the 1990s and beyond, as a greater proportion of their gas comes from non-European countries, namely the USSR, Algeria and elsewhere. Gasunie has made it clear that it is willing to act in this way and has the capability to do so. However, it has also been very clear about the fact that importers will have to pay a price for this valuable service.[35]

In certain circumstances, the UK might welcome the prospect of Dutch supplies, either as a peak load facility or in larger quantities, particularly as the Southern Basin fields become depleted. The logistical link between the Dutch and UK North Sea sectors would be short and might not require a great deal of additional pipeline capacity. In the early part of 1984, immediately following the agreement on the terms of the Sleipner contract, BGC was instructed by the Department of Energy to consider alternative purchases of Dutch gas.[36] Dutch willingness to provide such an alternative was announced publicly at a major British gathering of international energy economists.[37] In the event, BGC insisted that Dutch supplies were not competitive with Sleipner and negotiations were never seriously pursued. It remains to be seen whether the Dutch, having extended the contracts of existing customers for a further ten years, will be interested in attracting new customers, particularly if it requires investment in new infrastructure. There would need to be a decision that, in addition to the last round of extensions, yet more gas was available and that the Continental European importers were not in the market for further deliveries.

An obstacle to British imports of Dutch gas would be the cost of the link to the Southern Basin, which would undoubtedly have to be borne by the UK, and might not be considered worthwhile if the quantities were to be small. It is probable that any future imports from the Netherlands would be in the order of 5 BCM per year, rather than the larger quantities which other suppliers have the capacity to offer.

The USSR already supplies six West European countries with natural gas: West Germany (including West Berlin), Italy, Austria, Switzerland, Finland and France (in addition to six East European countries and Yugoslavia) through a pipeline system which stretches as far west as Paris. A link to the UK would require an extension of the pipeline network westward and a cross-Channel pipeline to the UK mainland, or a displacement agreement with the Netherlands (this will be discussed below). It is clear that the USSR, with 40% of the world's proven gas reserves, has a large exportable surplus of natural gas to sell to Western Europe and is keen to do so in order to earn hard currency.[38] The USSR would be in a position to supply very large quantities of gas to the UK within a comparatively short time. Given the spare capacity in existing pipeline systems, it would be possible for some 10 BCM of Soviet gas per year to be flowing to the UK within five years.[39] The only physical (as opposed to political) constraint is the speed at which the link to the Continent could be built. In theory Soviet gas could have been in competition with Sleipner gas for the UK market in the early 1990s, and purely on commercial grounds the USSR would be a strong competitor in any future UK choices between external sources of supply.

The UK market has a considerable attraction for the USSR, since it is the only major gas market in Western Europe (with the exception of the Netherlands) which it has not yet penetrated. For the UK, Soviet gas may be an attractive commercial proposition, both in terms of the competitiveness of imports and the possible export opportunities in the steel pipe and engineering industries. The Soviet authorities wish to sell additional large quantities of gas in Western Europe, but will be to some extent prevented from boosting their supplies to existing customers by concerns about security of supply, which were voiced in the transatlantic debate over the new Siberian pipeline. The UK market is sufficiently large for security concerns to be less acute; even if the USSR were to supply 10 BCM per year to the UK by the end of the century, this would not breach the limit of 30-35% set by the International Energy Agency regarding acceptable dependence on a single supplier.[40] Beyond the end of the century there would be scope for increasing imports from the USSR since there is every likelihood that sufficient Soviet export capacity will be made available in the longer term. The actual figure would depend on the capacity of the cross-Channel line; the decision on the dimensions of the facility would therefore be an important consideration.

However, it is not certain that Soviet gas would be politically acceptable, despite the prospect that it would be offered at competitive prices. The future of the transatlantic debate over the political wisdom of importing gas from the USSR may be an important factor in determining whether this source of energy is considered seriously in the UK. Certainly, up to the mid-1980s, this has not been the case. Soviet gas was not considered as an alternative to Norwegian Sleipner gas in the early 1990s, although it is likely that a comparable quantity of gas could have been available and the Soviet authorities would have been amenable if approached. Despite the fact that BGC has, as one would expect, considered the logistics of such an import, a BGC official admitted in front of a House of Lords Committee in early 1982:[41]

> I have to say that at present we have not directed our minds very closely to the idea of importing gas from Russia. As years go by, maybe that option will become more prominent in our thinking. However, I have to say that at the moment it is not prominent in our thinking. We tend to look firstly towards Norway and secondly - and, I think, less desirably - to LNG imports from further afield.

Indeed until relatively recently, neither BGC nor the Department of Energy could bring themselves to mention the USSR by name in official publications. The 1978 Green Paper was prepared to countenance pipeline imports from the Middle East, but the USSR was presumably included under the geographical heading of 'Asia' along with Afghanistan, the only other pipeline gas exporting country on that continent.[42] A Conservative Government could be expected to greet the idea of Soviet gas supplies with particular concern and, when rejecting the Sleipner import, the Minister made it clear that if and when imports were considered again, Norway and the Netherlands would be the only options of interest.[43] Other UK interests have expressed disquiet at the potential security implications for the UK of Soviet gas imports.[44]

By 1985, BGC thinking had obviously moved forward a quantum leap with the Chairman quoted as saying: [45]

> ..nobody has told me that I can't talk to the Russians...One doesn't have to talk to the Russians to know what their position is, but one might go to talk to them.

It remains to be seen whether BGC's political freedom of action would actually allow it to negotiate meaningfully with the USSR, as opposed to using the threat of going to Moscow as a negotiating card with other suppliers.

Other suppliers of pipeline gas to the UK in the future might be North and West African countries and, eventually, the Middle East and Gulf countries. A potential contributor to a cross-Channel pipeline might be Algeria, which in 1983 began to pipe natural gas across the Mediterranean to Italy. Feasibility studies have been carried out regarding the possibility of laying a pipeline across the Straits of Gibraltar to Spain. Given the possibilities of pipeline gas from Siberia to the UK, there is no logistical reason why Algerian gas should not travel the considerably shorter distance to the UK via Italy or Spain. The major obstacle at present is Algerian pricing policy and concern about the country's reliability as a supplier, which has complicated its gas trade with Italy and France.[46] Nevertheless, it would be wrong to rule out the long-term prospects for Algerian gas exports to the UK. The country's resource base and current export commitments would seem to indicate that a long-term export commitment of 5-10 BCM per year could be allotted to the UK. The lower end of the range would be more realistic, however, and it would therefore be reasonable to conclude that it would not be worthwhile to construct a pipeline solely for Algerian gas imports.

Similarly it is reasonable to consider imports of pipeline gas from as far away as the Middle East and West Africa. These sources are usually considered only in connection with trade in liquefied natural gas, but if a cross-Channel pipeline to the Continent were to be in existence, there would be an attraction in capitalising on as many sources of gas as possible. Early in the next century, Middle East gas may be considered more carefully by Continental European countries, given the vast volumes of natural gas in the region which appear to have no alternative large-scale domestic or export markets. Pipelines from the Middle East to Europe (via a number of possible routes) or from West Africa (mainly from Nigeria but also possibly Cameroon) via a Trans-Saharan pipeline from Nigeria to Algeria and thence to Europe are technically feasible, though the cost of transportation would be an important element affecting their economic viability.[47] A more important obstacle would be the number of countries the lines would have to cross, and this aspect would require detailed security analysis. The Continental countries are likely to lead in this endeavour, but the UK should certainly take an interest in receiving a share of the very large volumes which would need to be involved in order to make the projects economically viable.

A pipeline link to the Continent

Until 1980 there had been very little public discussion of the possibility of a pipeline link to the Continent. In the 1978 Green Paper it was noted:[48]

> If the UK and Norway decided to collaborate in the development of gas resources in the Northern Basin of the North Sea, it might then be feasible, if the Norwegians so wished, to re-export their portion of the gas across the Channel to the Continent. In the long term, when our supplies are in decline, a cross-Channel link would open up the possibility of the UK importing gas by pipeline from Asia or the Middle East.

A pipeline link to the Continent can be placed in a number of different commercial contexts, with different export and/or import alternatives. In addition, there is the prospect that the UK might play a significant role in an interconnected West European gas grid which might enhance security arrangements for the entire region.

Exports. During the 1970s, there was opposition from both BGC and successive Governments to the idea of a cross-Channel link, mainly because of objections to exports. Given the statutory responsibilities of BGC to the UK consumer, this is entirely understandable. Internationally traded gas is sold on long-term contracts and any trade which would warrant the building of a pipeline would involve rather large volumes. A contract which involved even 5 BCM per year for twenty years, would subtract 100 BCM from the proven reserves (nearly 14% of the present total) and would require careful consideration in terms of the national interest. It is equally understandable that the oil companies, which consider that they have always been paid too low a price for their gas by BGC, wish to gain access to a market which has a tradition of paying higher gas prices, particularly in cases where the pipeline link to the Continent would be easier than to the UK.

A BGC view expressed to a House of Lords Committee in mid-1982 was that:[49]

> We believe at the moment that the establishment of a pipeline link very clearly carries with it the implication of the export of indigenous gas to Europe and we express concern about that suggestion.

Such fears were heightened earlier in that year when the Secretary of State for Energy raised the prospect of exporting gas from the UKCS. At the same time as extolling the virtues of the Oil and Gas Enterprise Bill (1982), he added:[50]

> all the gas in our offshore fields currently in production is contracted to be sold in Britain, so the question (of exports) arises only for future fields ... If, however, the fresh impetus which our policies will undoubtedly give to exploration results in large volumes of new gas being discovered, the question of exports can and will be reconsidered then.

Mr Lawson did not make it clear how much gas would need to be discovered in order to raise the possibility of exports and whether he was thinking of large-scale quantities or marginal volumes. When considering the overall UK gas resource base, it would be difficult to predict the quantity of gas which would need to be discovered to allow exports of more than marginal volumes, while not endangering long-run security of supply. An example of the pitfalls which lie ahead in any such exercise can be seen in the case of the Netherlands, where large-scale exports were contracted in the 1960s, only for the jolt of the 1973/4 oil crisis to bring the country to contract for imports of Norwegian gas and then subsequently in the 1980s to reverse its policy again by raising the prospect of increased exports. This provides an example of how a country, with a smaller domestic consumption than the UK and more than twice the proven reserve base, can be panicked into taking illogical action when its long-term security of supply appears to be threatened.

The issue of exports came once again very much to the fore in the mid-1980s, in the context of the proposed import of Sleipner gas. The position of most oil companies on Sleipner was that they did not oppose the import (or indeed the concept of imports) but they believed that this should be accompanied by freedom to export gas from the UKCS. In fact, present legislation does not prohibit exports of UKCS gas. However, the constraint of 'the landing requirement' (defined in the 1966 Petroleum Regulations) means that the Government is required to give consent for the gas to be landed elsewhere than in the UK.[51] The problem with most comments on the concept of exports is that they are made, either as part of a negotiating position between BGC and producing companies regarding the timing and pricing of resources to be developed, or as part of a statement of economic preference in favour of a 'free market' in natural gas. The main proponent of the free market argument is Colin Robinson:[52]

> I would contend also that the size of proven reserves is not independent of the foreign trade and market regimes. Proven reserves of gas would now almost certainly be higher if Southern Basin exploration had not been discouraged for so long by BGC's monopsony and monopoly position. Very probably freer trade in gas would give a considerable stimulus to exploration and development so that more reserves would be proved. The seductive argument that exports cannot be allowed because proven

reserves are insufficient is thus unconvincing; proven reserves never will appear sufficient unless there is freedom to trade.

This argument may have merits over a limited time frame, but is flawed in the sense that, while incentives can be provided for more exploration, this is only useful if the geological resources actually exist. If the resource base is limited within the boundaries we currently believe, encouragement to explore, by means of allowing exports, simply means that indigenous resources are depleted at a faster pace. It is only if we place Robinson's views alongside the larger resource estimates in Table 2, that we get the excellent prognosis that the more gas we produce and export, the more gas we shall find.

The notion that exports are important to maintain incentives for exploration is often expressed. From a producer point of view, there is a desire to maintain expertise within this sphere of activity, not simply on the exploration and production side, but also for negotiation and selling functions. If operations have been run down to a low level, this type of expertise cannot suddenly be rediscovered in response to a change in conditions. A common oil company position on exports is that, if BGC is not prepared to buy the gas at an appropriate price, producers should be free to find an alternative purchaser. A more sophisticated version of this position was put forward by BP during the Sleipner inquiry:[53]

> To justify leaving gas in the ground as an investment the nation must believe that the value of the gas will increase at a higher rate than the rate of return from investing the revenues realised from the gas development. There is considerable uncertainty about future levels of international energy prices but we believe that these are likely to increase by at most 3% per annum in real terms to the end of the century. We consider that the nation should be looking for and be able to achieve a better return than this on its investments...there will be occasions when the UK does not value gas in its own system as highly as other people value it. It would be in no one's interest whatsoever that we consume gas in that way, or that we allow it to lie fallow in the field. It would be much better in those circumstances to allow it to be exported and to invest the revenues.

This was a restatement of BP's view on oil to the 1981/82 Select Committee inquiry on oil depletion policy. In their final report on oil depletion policy, the Committee supported BP's view, but they did not directly do so in their gas report, perhaps taking the view that gas exports would have a much larger effect on the resource base, owing to the need for large-scale commitment of resources to long-term contracts.[54]

BGC's argument is that it is prepared to buy all gas discovered on the UKCS and that it is currently offering prices which provide an entirely adequate incentive both to explore and to sell to the Corporation. While it is extremely difficult for independent observers to verify what is being demanded by producers and offered by BGC, in the mid-1980s there appeared to be a queue of producers waiting to negotiate with the Corporation for gas sales. This may reflect a short-term overcommitment of the Corporation in the area of gas purchases in the period up to 1991. In the 'post-Frigg' period, however, especially after 1993, the picture is markedly different. Questions to the interested parties about this situation invariably bring forth a statement of negotiating positions on prices.

In the midst of these negotiating positions it is possible to draw one main conclusion: nobody currently operating on the UKCS is keen to export large quantities of gas to the Continent. There are very few occasions when a proposal to export UKCS gas has reached the point of including details of a field, a quantity, and a purchaser. While potential exporters might argue that there is no point in putting forward such propositions unless there is a genuine expectation of obtaining permission to export, the more reasonable conclusion is that producers wish to use the threat of exports to force BGC to pay higher prices. An intriguing question in the mid-1980's is the size of the market which exists on the Continent for aspiring exporters of UKCS gas.

The only exception to these general observations would be co-operation with Norway in the exploitation of the associated gas in fields which are close to, or straddling, the sector line. For example, it is evident that UK Statfjord gas could have been more easily transported through the Statpipe system to the Continent (once the decision had been taken to send the gas from the Norwegian sector to that destination) than to the UK mainland, and this was discussed in the context of the Sleipner negotiations. For a number of reasons, outlined above, this did not happen and, with hindsight, this was almost certainly a missed opportunity to negotiate a quid pro quo for base load Norwegian supplies.

Imports, re-exports, displacement agreements and security. The second area where a pipeline link to the Continent should be seriously considered brings in the concept of West European gas security and the future involvement of the UK in the Continental gas grid. There is no counterpart in natural gas trade to UK participation in the IEA emergency oil sharing mechanism, or the cross-Channel electricity link (with France) which will start in 1986.

It would be of considerable benefit to Continental countries to have the assurance of UK supplies for emergency contingencies. However, it is difficult to imagine that a pipeline to the Continent, the sole purpose of which was to permit occasional gas transfers for security reasons, would be possible to finance commercially or of any concrete benefit to the UK. To the sceptical, this would still seem a thinly disguised excuse for exports of UK gas on a more than marginal scale.

By contrast, it is perfectly possible to envisage a pipeline, the primary purpose of which is base load imports, but which provides for exports in emergency situations. The two obvious examples would be imports of Dutch gas via the UK Southern Basin pipelines, and of Soviet and/or Algerian gas through a cross-Channel pipeline from either France or Belgium. Another would be the re-export of Norwegian gas transitting the UK, which might involve an export of UK gas from the Southern Basin as a quid pro quo for additional imports of Norwegian gas. The proposal to re-export Norwegian gas to the Continent was put forward at the time of the Statfjord negotiations.

Yet another example would involve the UK importing Soviet gas in a displacement agreement with the Netherlands. In this scenario, although the contract would be signed between BGC and Soyuzgazexport, the actual gas received would be Dutch; Continental countries would receive additional supplies of Soviet gas.[55] With the necessary adjustments in transmission capacity, the Netherlands would retain the ability to switch supplies to Continental countries if an emergency were to arise; additional UK gas might then be required to reinforce Dutch supplies. In each case an additional requirement would be the ability to reverse the flow in the pipeline connecting the Dutch sector to the UKCS when security dictated. In its annual reports, the International Energy Agency has repeatedly urged the UK to:[56]

> Evaluate the consequences of waiving the landing requirement for natural gas and evaluate the benefits and viability of connection to the natural gas production and distribution system on the European continent, in light of the security benefits to the UK and Europe of the potential flows of Norwegian and Continental supplies.

One of the main impediments to a pipeline link between the UK and the Continent has been hesitation on the part of BGC and the Continental utilities on the question of who would control the flow of gas. There has been concern from each side that, at a time of supply crisis, the utility at the end of the pipeline would be the one to suffer shortages. Such scenarios depend on the degree of trust which exists between the parties, but the arrangements currently in place on the MEGAL and TENP pipeline systems suggest that there is a considerable advantage in belonging to a very large pipeline grid in that problems which arise can be shared out with minimal disruption to any one utility.[57]

However, irrespective of the technical arrangements, BGC would certainly need the goodwill of Continental utilities, if gas destined for the UK were passing through their service areas or through their facilities. The adversarial relationship between BGC and the Continental utilities in bidding against each other for the same sources of supply is likely to be of less importance in the 1990s owing to the relative abundance of potential imports from Norway and the USSR. It could be advantageous for all parties if this were to be replaced by joint co-operation in expanding flexibility of the total West European gas grid. A pipeline link between the UK and the Continent would be a considerable advance in this regard. In the mid-1980s, concern about the security of West European gas supplies had abated somewhat from the fever pitch of the early part of the decade; however, this issue will undoubtedly be raised again and will remain relevant to the future debate on sources of supply.

LNG options

The first commercial liquefied natural gas trade was initiated between the UK and Algeria in 1964. A regasification terminal was constructed at Canvey Island and this handled about 1 BCM of LNG per year from the Arzew 1 liquefaction plant until the contract expired in 1980. A supplementary nine-month agreement expired in October 1981.[58] Since that time no LNG has been received at Canvey Island, partly because of an accident which severely damaged the jetty in May 1982. This would in any case have precluded the berthing of LNG tankers, but there has been no suggestion of restarting trade with Algeria in the interim period; the jetty was repaired by the autumn of 1983.

By modern standards, the Canvey Island terminal is extremely small and can only be considered as a 'peak load' facility. Quite different facilities would be required if LNG supplies were to take over a 'base load' role in the UK. The most recent LNG terminals built in France and Japan have a capacity of 7-12 BCM per year and the UK would require 2-4 terminals of this size to meet 50% of present demand.[59] Furthermore, the lead times for constructing a modern LNG terminal are very considerable. BGC estimates the construction time for a terminal as:[60]

> at least five years, depending on the amount of site preparation necessary. The award date would be preceded by a period of time during which conceptual designs would be prepared, including hazard analysis and environmental impact analysis studies and planning applications made. The minimum time for this activity would be two years. However, this preconstruction period could easily be extended if a public inquiry was to be held and an extended period of public consultation necessary. In this event, the preconstruction period would last 4-6 years. Thus the minimum time needed to bring an LNG terminal to the operational stage from initial nomination of a specific site would be 7 years, but this could be extended to over 10 years.

In preparing this estimate, BGC may have in mind not simply the experience of other countries, but the potential opposition from local residents which, in the case of Canvey Island, has led to three reports on the safety aspects of the facility in the period 1978-83.[61] The third report concluded that, after a period of unsatisfactory performance in the late 1970s, the terminal could be considered safe and that capacity could be maintained at current levels. However, BGC gave an undertaking that the capacity of the terminal would not be expanded. Thus new sites would be required in order that large-scale LNG supplies could be imported. Using the time frame outlined above, this would preclude any large-scale import of LNG before the mid-1990s, even if a terminal were under consideration by BGC; there is no present indication that this is the case.

As regards sources of supply, Algeria would certainly be the prime source of LNG (presupposing that this was considered preferable to the pipeline alternative outlined earlier). Other possible sources of LNG would include Nigeria, (Arctic) Canada, Trinidad, Cameroon and the Gulf (principally Qatar). On present knowledge, it would be unrealistic to assume that Canada and Trinidad would yield more than marginal volumes of gas in any time frame. For domestic economic and political reasons neither Nigeria nor Qatar is likely to develop a transportation infrastructure which would yield large quantities of gas before 2000, quite apart from the question of UK readiness to accept such supplies.[62]

After 2000 and certainly prior to 2020, it is perfectly feasible that the Middle East countries, and also Nigeria, could be sources of considerable volumes of LNG (assuming they had rejected the pipeline alternative). However, there are major uncertainties, particularly with regard to the route for the Middle East trade; it is not certain that large numbers of tankers passing through the Suez Canal would be environmentally acceptable, and the route around the Cape would be enormously costly.

Overall, it has to be said that, as for so many gas-importing countries, the LNG option (pioneered by the British two decades ago) looks much less attractive in the 1980s than most of the pipeline import prospects outlined above.

Substitute Natural Gas

Apart from natural gas from the UKCS and the various import options, there is the prospect of producing gas from other fuels - mainly coal and oil. This is generally known as 'substitute' natural gas or SNG, and there are a number of technologies which are at different stages of development. At the present time, there is no large-scale high calorific gasification programme in existence, and in the mid-1980s plans for starting such programmes have been shelved worldwide because of falling oil prices. The only consideration here will be high calorific value gasification, i.e. gas which is interchangeable with natural gas in the national grid, and specifically, gasification of coal. There are also options to produce lower calorific value gas from coal, which can be used in combined cycle power plants or as town gas. As far as gasification of oil and oil products (heavy oil and naphtha) is concerned, although the UK may run short of natural gas from the continental shelf prior to exhausting its oil reserves, the cost of production and the international price of oil are likely to preclude large-scale conversion from oil to gas. Even so, the two plants at Portsmouth and Plymouth which use naphtha as feedstock are an indication that BGC is interested in this prospect. One concern must be to avoid the conversion of any part of the gas grid to lower calorific value gas. Having made the transition to natural gas with a massive conversion effort in the late 1960s and early 1970s, it would be inconvenient (as well as expensive) to reverse that process in the future.

BGC has been conducting a long-term research programme of coal gasification at Westfield in Scotland. By 1981, the BGC/Lurgi slagging gasifier had completed 90 days of extended trials during which 1,750 cubic feet (approximately 50 cubic metres) was gasified from 27,000 tons of coal at a rate of 300 tons per day.[63] Commercial SNG plants are likely to produce 2.5 BCM per year, requiring an average of 5 mt of coal per year (this is an approximate figure which ranges from 11,500 tons per day of high quality coal to 25,000 tons per day for poor quality lignite). At present, 20-25 plants would be needed to satisfy the whole of UK gas demand, and this would require 100 mt of coal per annum. The approximate capital cost is thought to be around £1 bn per plant and the cost of the gas is estimated to be 'two to three times as expensive as natural gas', or 'upwards of $50 per barrel (1982 dollars)'.[64]

The sites for SNG plants require access to large quantities of coal and water (50,000 cubic metres per day) and disposal facilities for solid waste. A commercial gasifier is estimated to produce 2,000 tons of slag, 230 tons of sulphur and 5-10 tons of sludge on a daily basis.[65] The Commission on Energy and the Environment suggested that there was a need for an intermediate demonstration plant of 0.5 BCM per year capacity, prior to the building of full-scale 2.5 BCM per year plants.[66] BGC estimates that:[67]

> from the date of award of the design and construction contract to satisfactory operation, a period of 7 years would elapse. As in the case of any major industrial site ... the preconstruction period could be considerably extended by the requirements of public planning inquiries. The whole process could therefore well extend to 10-14 years. Based on the construction period alone, it is estimated that a period of up to 10 years would be needed to create 3,000 million therms (about 8.3 BCM) of SNG and up to 15 years to create 6,000/9,000 million therms (16.5-25.0 BCM) of SNG capacity, following the successful completion of a proto-type plant. From about 1985, when further development work has been completed, it would be possible to embark upon a construction programme which would give SNG by the mid-1990s if this were essential. This would imply a bypassing or modification of the prototype programme and could give rise to increased costs of construction and operation.

A report in 1983 that BGC was intending to build 20 coal gasification plants, starting in 1990, would therefore imply an accelerated programme, either speeding up or eliminating the prototype stage.[68] Even starting in 1990, the first plants would probably not be ready much before 2000-2005 and it seems likely that one would be built and tested before a full-scale commitment of funds was made. Alternatively, construction and testing of a prototype plant might push the first commercial station back to the second decade of the next century. In any case it would seem prudent not to rely on large-scale supplies of SNG before this date. The only development which could radically change these judgements would be a technological advance in the United States, leading to rapid commercialisation, which could be imported by this country on a large scale. While this should not be ruled out, the current slump in both interest in, and funding for, synthetic energy technologies in the United States does not give cause for optimism.

Overall, it is clear that the lead times and the technical uncertainties are such that to produce appreciable quantities of SNG prior to 2000 would require a very large commitment of funds and would also envisage moving ahead without a properly tested technology at the prototype stage. In the 1980s, the gas supply and availability picture is moving further and further away from any large-scale consideration of SNG supplies until several decades into the next century. The availability and price advantage of natural gas, both domestic and imported, should give rise to consideration of whether it is even worth continuing to put substantial research funds into SNG technology.

The only conceivable reason why SNG supplies might be brought forward as a large-scale source of supply prior to 2020 would be the ruling out of any imports of natural gas, while maintaining the UK gas market at its present size. SNG from coal provides the only domestically produced gas available after the depletion of UKCS supplies and is therefore crucial to any aspiration towards natural gas self-sufficiency. (It would, of course, be possible to base an SNG programme on imported coal, although it is easy to envisage tremendous resistance from the British coal industry to such action, and the self-sufficiency advantage is immediately negated.) The price of self-sufficiency using SNG supplies would be enormously high, perhaps twice as much per therm as domestic or imported supplies, and it would be extremely difficult to think of a set of circumstances in which this option would become attractive, prior to 2020.

Natural Gas Pricing

Perhaps more than any other issue pertaining to natural gas supplies, pricing is the area where views are strongest and differences of opinion widest. The pricing of natural gas has two aspects: the price which BGC (and other potential purchasers) are willing to offer producers (and exporters) - generally referred to as a 'wellhead' or a 'landed' (beach) price; and the price paid by the UK consumer - domestic, commercial and industrial - for gas supplies. The magnitude of both and the relationship that the consumer price should bear to the landed price are matters of continuing, and at times angry, debate.

Producer prices

Unlike the situation in the international oil trade, there are no commonly accepted international prices for natural gas. Another major difference from the international oil business is that natural gas prices in Western Europe - both the actual figures and the way in which they are arrived at - are confidential. In the UK this applies to prices that BGC pays for both UKCS and imported supplies.

While information on West European pricing has improved greatly in the 1980s, following on the upsurge of interest in the subject, it is exceedingly difficult to obtain anything other than a confidential, word-of-mouth estimation of a price at a particular point in time. Moreover, figures, without the explanation of their context, are not particularly helpful. In a world of volatile international energy prices and exchange rate movements, it is the details of escalation clauses and currency arrangements that provide the key to how prices might move over time. This has become particularly complex in the mid-1980s as the prices of imported gas into Continental Europe have become increasingly geared to the prices of competing fuels in end-use markets.

UKCS prices. In the UK, BGC was, by law, both a monopoly purchaser and sole vendor of natural gas until the passage of the Oil and Gas Enterprise Bill (1982). By the end of 1985, this legislation had produced no effective change in BGC's monopoly position.[69] The central argument between the producer companies and BGC concerns the price which the latter has been willing to offer for supplies from the UKCS. The view of the companies has been that, throughout the 1970s, gas prices were too low to stimulate exploration and production.[70] These complaints combine with resentment at the much higher price that BGC has paid for supplies of Norwegian Frigg gas. In the 1980s, there has been a recognition that BGC is willing to pay 'higher' prices and, although there is grumbling from the companies that the levels are still not high enough, there are indications that gas prices in newly negotiated contracts are in the range of 20-26 pence per therm. However, it must be stressed that, as far as prices are concerned, simply quoting a number and comparing it with another number is meaningless without knowing the volume, delivery profile, escalation clause, and currency of payment, plus a very large number of other variables.

It is certainly true that BGC has paid - and has offered - higher prices for gas imports than to UKCS producers. During the financial year 1983/84, the Norwegian Frigg price was around 24 pence per therm - 7 pence higher than the average cost of all gas (excluding the gas levy).[71] The reported base price offered for Norwegian Sleipner gas in early 1984 was around $4 per mmbtu (at that time roughly equivalent to 27 pence per therm), priced in US dollars and escalating with a mix of fuel oils. By the end of the year, the weakness of sterling had raised the price to 35 pence per therm, so that Sleipner had become 'expensive'. One year later, the strength of sterling combined with falling oil prices meant that Sleipner would have been around 25 pence per therm, roughly in line with new UKCS offers.[72] Had the import been approved, the pendulum would undoubtedly have swung backwards and forwards several times prior to the commencement of the contract in 1992. Moreover, it is unwise to insist on comparing an import contract of 10 BCM per year with a UKCS development typically yielding a much smaller annual volume.

Irrespective of parochial arguments about the fairness of BGC's past pricing policies vis a vis the producers, the inescapable trend is towards higher field development costs and hence the need to offer higher real prices for UKCS gas. Indeed in the 1980s, the real cost of UKCS gas may be rising rather faster than the cost of potential imports, although domestic gas started from a rather lower base. These trends are partly illustrated in Table 9, which shows that the average price that BGC paid for gas and gas feedstocks has risen steadily and swiftly over the past decade, the most dramatic increase being in 1984/85 when average gas costs rose by 3p per therm or 22.6%. Also, in 1984, the cost of gas (excluding the gas levy) exceeded operating costs for the first time.

From fragmentary data, it can be calculated that the average price paid for UKCS gas in 1984/85 was roughly 12 pence per therm which, contrasted with present levels for new contracts of 20-25 pence, indicates the large proportion of old contracts still included in the total.[73] It is quite clear that new UKCS supplies from associated gas and condensate fields will cost considerably more to produce, compared with the relatively large dry gas fields of the Southern Basin, and prices for UKCS gas are likely to rise accordingly, particularly in the last decade of the century as the original Southern Basin fields go into rapid decline.

Import prices. However, if the costs and prices of UKCS gas are expected to rise sharply, the cost of the majority of available imported supplies is hardly likely to provide a much more attractive alternative. On a rough calculation, the average price of Frigg gas in 1984 had risen to around 28 pence per therm, contrasting with the agreed Sleipner price fluctuating wildly around 25-35 pence in 1984/85.[74] This gives an indication as to the rough price of base load supplies from Norway. A further indication will be given by the price for supplies from the Troll field (Phase 1) development. The parties have indicated that a preliminary agreement should be signed in 1986 and this will set the scene for Norwegian gas prices in the 1990s and the early part of the next century.[75]

Looking at other base load sources of supply, the question of the price of possible Soviet gas supplies may be extremely important in the 1990s. According to press reports, the price for long-term contract Soviet gas delivered to the Continental utilities in late 1985 was in the range of $3.50-4.00 per mmbtu, mostly denominated in the currency of the importing country.[76] Considering the possibility of a dedicated pipeline to the UK, one might, for illustrative purposes, take the French price and add a notional transportation cost for 600-700 km of pipe for the extension of the grid across France (and possibly Belgium), including a Channel crossing of up to 50 km in water depths of less than 200 metres, of 50-60c per mmbtu, bringing the c.i.f. price of Soviet gas on the UK mainland into the range of $4.00-4.60 per mmbtu (roughly equivalent to 25-30 pence per therm at late 1985 exchange rates). Much might depend on whether the USSR would be prepared to make concessions on the price to the UK in recognition of the higher transportation element and, perhaps as critical, whether the Russians would agree to a sterling price.

As mentioned above, there is an incentive for the USSR to be conciliatory to the UK over gas sales, given its desire to increase gas exports and hard currency revenues and the market limitations in the Continental countries. If, therefore, the Soviet authorities wish to sell additional large quantities of the fuel in Western Europe, their best prospect will be to approach the UK with a large quantity of gas at an attractive price. It is, of course, easier for a non-market economy to manipulate the price of an export commodity, and while it is unlikely that the USSR will attempt greatly to undercut the 'market price', it is certain that Soviet gas will be priced very competitively against all of its alternatives.

The question of Algerian gas pricing, either for LNG or pipeline gas, is rather more complex. As noted above, the UK imported Algerian gas for 17 years in what proved to be a satisfactory and profitable trade for both parties. The contract expired in 1980, but there was a nine-month extension under which volumes were delivered at $4.60 per mmbtu f.o.b. rising to $4.80 per mmbtu f.o.b. in July 1981.[77] The parties were then unable to agree a price for further deliveries and the trade ceased at the end of 1981.[78] The inability to agree price terms for a further extension was linked to Algerian demands for parity with crude oil in the base price and escalation clauses in any new contract. This was a demand that BGC was unwilling to meet, even for a peak load contract where the capital investment had been amortised.[79] The prospects of BGC being prepared to sign future base load contracts with Algeria for LNG or pipeline gas will be related to Algerian performance in its trades with Continental countries, mainly Italy (pipeline gas) and France (LNG). If Algeria continues its past attempts to enforce price and volume conditions considered unrealistic by buyers, there is little likelihood that the UK will be interested in purchasing its gas. Any indication of a change of heart either in London or Algiers might be reflected in some resumption of small-scale LNG trade through the Canvey Island facility.

High price demands, combined with previous instances where Algeria has demanded price renegotiation and curtailed gas deliveries in support of its price demands (in the case of the Continental contracts), would deter BGC from entering into base load supply contracts with the country in the immediate future.[80] However, it is worth repeating that, in the case of the original BGC contract, the experience over a long period of years was extremely favourable and that, given a change in Algerian government policy and a number of years of satisfactory trading experience with Continental countries at acceptable prices, there is no reason why Algeria should not be reconsidered as a supplier of LNG or pipeline gas.

Prices for possible gas deliveries from the Netherlands will depend on whether peak or base load supplies are being contracted. However, it is likely that the UK could expect the same terms as Continental countries, with a small additional allowance for new pipeline links which would need to be constructed. While the cost of Dutch gas production and transportation will be low compared with other sources of supply, the Netherlands Government and Gasunie will expect the price to reflect the logistical and political attractiveness of Dutch imports.

All other base load supplies would be more expensive, in terms of their cost of production and transportation. Despite initial optimistic forecasts of LNG availability and cost in the mid-1960s, when it was believed that 'imported gas could be landed on these shores from as far away as Nigeria and Venezuela for about $4\frac{1}{2}$d per therm', LNG costs have escalated dramatically, in comparison with the cost of pipelining gas.[81] One large-scale, long-term option, i.e. beyond the turn of the century, would be pipeline imports from the Middle East, where very large quantities could be made available on a scale that might reduce the price per unit to an acceptable level, particularly if the facilities were being shared by Continental utilities.

Table 10, which summarises some illustrative costs and prices for domestic and imported gas, raises some essential questions for future supplies and the choices between available alternatives. First, it suggests that incremental imports from any source will be between 50% and 100% more expensive than the current average cost of UK supplies and roughly two to three times the average price currently being paid for UKCS supplies. Second, if the illustrations are roughly correct it suggests that, on price grounds, there may be very little to choose between Norwegian and Soviet supplies. It is a great pity that, at the time of writing, there is no indication of a price being discussed in the Troll negotiations between Norway and the Continental consortium. This will certainly constitute a benchmark for future Norwegian export prices. It should be stressed that the figures in Table 10 are illustrative, in the sense that much will depend on delivery profiles, exchange rates, and the cost of the transmission system.

Consumer prices

Consumer prices are the most directly politically sensitive area in the UK gas industry. The government is aware that BGC's 16 million customers all have votes (and most households have more than one vote). Public opinion has proved to be extremely sensitive on the question of whether counterparts (or competitors) in other European countries are paying more or less for their gas than is the case in the UK (although this is usually difficult to calculate and changes rapidly over time owing to exchange-rate fluctuations). There is a sizeable slice of public opinion which holds that it is right and proper for UK citizens to pay less for 'our gas' than less fortunate Continental countries which have to import supplies. Such a view takes little cognisance of the effects of the price mechanism on conservation, or the rising cost of future supplies. Emphasis is often placed on the privations suffered by poorer sections of the community which are disproportionately affected by rising gas prices. This is an extremely important social policy question which should be the responsibility primarily of the Department of Health and Social Security rather than the Department of Energy, although BGC has become increasingly involved in the problem in recent years.[82]

The two major features of consumer gas prices are, first, that following the first major oil price rise of 1973/74, they have generally been cheaper (particularly in the residential sector) than competing fuels and, secondly, that they have repeatedly been manipulated by successive governments for political and economic (mainly revenue-raising) reasons, which have little to do with energy policy. Vickers and Yarrow point out that:[83]

> priorities have been largely determined by macro-economic policy. Thus, at different times, there have been ministerial interventions to hold down prices during periods when prices and incomes policies were operative; force up prices when public sector borrowing targets became the dominant consideration; speed up investment programmes and slow down plant closures when unemployment was perceived as a major problem.

Table 11 showing average prices to domestic and industrial customers underlines the uneven pattern of increases over the past decade. For comparison, the dates of real price increases in the period since 1979 are also shown. The Deloitte Haskins and Sells Report of 1983 gives the details of these policy shifts:[84]

> a) constraint on prices which mainly affected domestic prices during the period when the Price Commission operated (1973-9);
> b) constraint on domestic prices in the autumn of 1979;
> c) the sequence of three domestic tariff increases of 10% per annum completed in 1982;
> d) the freezes on industrial and commercial prices during the period since 1980/81.

> Under these influences, average revenue per therm in real terms behaved as follows in the period 1975/76 to 1981/82;
> a) domestic tariff fell by 20%, then recovered most of the fall;
> b) non-domestic tariff approximately retained its level;
> c) commercial firm contract increased by 60%;
> d) industrial firm contract increased by 100%.

> Over the same period the cost of gas increased in real terms by 200% (150% excluding gas levy), and gross profit per therm declined in real terms by almost 20%.

During this period, the emphasis of government policy on pricing and investment for nationalised industries changed. In 1967, the Nationalised Industries White Paper stated that, wherever possible, prices should be set equal to marginal costs, with the emphasis on long-run, rather than short-run, costs. The 1978 White Paper switched the emphasis from price to the financial target (to promote cost efficiency), supplemented by external financing limits and a required rate of return on new investment programmes.[85]

A number of competing theories of pricing have been put forward by academic and business economists. In an earlier version of this study, I identified two major schools of thought on the subject of consumer pricing. At one extreme, there is the view that natural gas prices should be set by

28

the market, with supply and demand determining the level of prices and gas competing with other fuels in the market place. At the other, there is the suggestion that the price of gas to the consumer should, under present policy, bear a relationship to the expected long-run marginal landed price. Reference to the long-run marginal landed price is not so much an analytical economic concept as an attempt to identify the cost of the replacement (as opposed to the additional) therm, whether imported or domestically produced.

Newbery, in an excellent paper on energy pricing policy, comments on my two schools of thought: 'Both are relevant; the replacement cost should be interpreted as anchoring the future price, and the present price must clearly equate supply and demand for efficiency.'[86] Yet while Newbery sets out the theoretical principles with admirable clarity, any attempt to put values on variables such as short- and long-run marginal costs (SRMC and LRMC) is fraught with difficulty. Partly, this is a consequence of the confidentiality of contracts and the difficulty of projecting how producer prices for domestic and imported gas might be affected under certain scenarios of world oil and energy prices. However, equally difficult is the selection of a particular source of supply as a representative marginal (replacement) source, in either the short or long term. Arbitrariness is inevitable in both the choice of source and in the calculation of the future cost (and/or price) of that source.

In testimony to the Select Committee, BGC weighed in mightily against using LRMC as the major yardstick for gas prices, but admitted that the import price was the best single indicator:[87]

> (LRMC) can never be other than one of a number of factors to be taken into account in setting prices...The cost of purchasing new supplies is the largest element in the LRMC estimates...Norwegian Frigg has been used as an indication of the marginal gas purchase cost, because the supply was contracted in competition with other countries and the price is escalated in line with world and UK energy prices.

Having considered evidence from all the nationalised industries and the relevant Government Departments, the Committee was 'struck by the way in which witnesses, whilst stressing the belief that LRMCs contribute to efficient resource allocation within the economy, were reluctant to quantify them in anything other than vague terms'.[88] Researchers from the Institute of Fiscal Studies conclude that 'The estimation of LRMC cannot be precise; but this does not mean that the concept is inappropriate as a criterion'.[89] Yet if LRMC cannot be estimated with reasonable precision, its use in determining the level of prices must be limited.

A number of considerations should be borne in mind when considering consumer gas pricing. First, there is a very imperfect market place for energy; domestic consumers, who constitute 50% of the market, cannot switch from one fuel to another with impunity. Secondly, if the price of natural gas had risen to match that of oil during the two price shocks of 1973/74 and 1978/79, the domestic hardship and political repercussions would have been enormous.

Third, costs are divided into two categories: gas costs and operating costs (see Table 9). Up to the mid-1980s, as far as gas costs were concerned, average costs were less than marginal costs. However, looking at operating costs, the situation has been the reverse as a result of high investments in the grid during the late 1960s and early 1970s, which are amortised over a period of years. There has therefore been a tendency for the two components to cancel each other out.

Overall, it is difficult to support Deloitte Haskins and Sells in their argument that consumer prices should reflect marginal costs, which they define as the cost of Norwegian Frigg gas plus a 5 per cent real rate of return.[90] In the longer run, there is no reason to select Norwegian Frigg gas rather than Sleipner, Troll, Soviet, Algerian or indeed new UKCS gas supplies as the marginal source, or 5 per cent as the correct rate of return. And in the short run, consumer prices based on the cost of any one source of gas which is only a small part of total gas supply (20-25% in the case of Norwegian Frigg) will be grossly distorting and will inevitably involve major discontinuities when a 'replacement' marginal source has to be chosen. A Department of Energy official made the following observation to the House of Commons Public Accounts Committee, regarding LRMC in relation to the gas industry:[91]

> World-wide in gas utilities there is a practice of averaging, relating to a situation in which the supply of gas which a utility is selling will commonly have a number of components, brought in at different points in the past at different price levels. In fact, what gas utilities tend to do in a long period of rising prices, is to pass on to the consumer some of the benefit of the earlier, cheaper gas; that is happening in the UK now and I believe it is happening in other countries as well.

Gas pricing policy in the 1980s appears to have aimed at this compromise (although, as outlined above, actual policy has been subject to short-term political and economic imperatives). The aim has been to raise the price of gas to a level which reflects both marginal cost, defined very broadly with a large number of inputs averaged in, and the market price, defined so that the price of gas should not be greatly out of line with that of oil products. Furthermore, there has been an attempt to raise the price in small steps, rather than large leaps, so as to minimise both hardship to consumers and domestic political reaction.

A major recommendation of the Deloitte report was that a long-term pricing policy should be agreed between BGC and the Government (despite the rather slim likelihood that any such policy could be agreed upon, given the record of deliberate government intervention in the industry for mainly political reasons).[92] Interestingly, something of this kind has been devised by the Government in preparation for the privatisation of BGC. After the privatisation of the Corporation, industrial customers will continue to negotiate their rates directly with the gas supplier. Prices to tariff customers (i.e. those residential and commercial customers using less than 25,000 therms per year) will be supervised by the Director of Gas Supply, together with the Secretary of State, and determined by means of a formula shown in Table 12.[93]

> The formula contains two major variables, X - representing the efficiency factor or the required percentage reduction in operating costs per annum (which is supposed to remain at the same level for five years) and Y - representing the increase in the average cost of gas taken during the relevant year. These are respectively subtracted from and added to the RPI percentage change for the relevant year to give the Maximum Average Price per therm. Where X is greater than Y, the real cost of gas will be falling.

30

There are safeguards in the licence which give the regulatory agency the ability to scrutinise any decision by the utility to raise prices (or hold them down) by more (or less) than a certain percentage annually.[94] At the time of writing, it is still too early to judge how all of these intentions will work out in practice, with major elements of both the price formula and the regulatory structure and procedures yet to be determined. There is enough time for changes in the political landscape to change both the intention to privatise and the shape that such privatisation might take. However, the experience of privatised utilities in other countries suggests that simple formulae rarely suffice for complicated situations and rarely last for long periods before they are replaced by other, more complicated formulae.[95]

Intense discussion on what constitutes an 'efficient' price has taken place because of the belief that 'if the objectives of BGC, notably pricing policy, are clearly defined, a separate depletion policy is unnecessary'.[96] The conclusion that any kind of controlled depletion policy is irrelevant would be a distinct change from the policy which BGC has adopted up to the present, which places considerable limitations on gas sold to non-premium markets and refuses to countenance exports.

Finally, it is worth stressing three long-running obstacles to any serious discussion of gas pricing policy. First, the problem that natural gas contract prices, both in the North Sea and Western Europe in general, are shrouded in confidentiality; it is high time that a greater degree of transparency was accepted by all parties and this could be one of the major contributions of the new regulatory authority.[97] Secondly, there must be doubts that future (costs and) prices of domestic or imported gas can be projected with any degree of confidence (although it may be easier for imports than for UKCS supplies). Thirdly, and probably most important, there seems little hope of achieving consistency in future consumer pricing policy. It might be a comforting thought that privatisation and the setting of a price formula will act as a stabilising force, but in practice there is a very long way to go before Government can be prevented from pursuing short-term political and revenue-raising objectives, in preference to energy/gas pricing considerations.

The Demand Situation

Tables 13 and 14 show the position of natural gas demand within the total energy balance, and the sectoral breakdown of natural gas consumption. Natural gas demand in the 1980s has held up well at a time of static or falling energy demand and sharp increases in prices, slowly but steadily increasing its share of primary fuel consumption to around 23%. The domestic sector dominates all others, with more than half the gas sold in the country. This accounts for the relatively sharp seasonal peak in demand requiring sophisticated transmission and storage infrastructure. While nearly one third of demand is in the industrial sector, there is virtually no gas used for power generation, unlike the situation in many Continental European countries.

Tables 15 and 16 are taken from the evidence to the House of Commons Select Committee and show ten estimates of gas demand in 1990 and 2000, made by Government, nationalised industries and producers. It is worth remembering that the assumptions and methodology behind these projections will not be identical and these factors may account for some of the divergencies in the estimates of 48-57 BCM in 1990, and 40-65 BCM in 2000, with the average of the projections indicating a demand level of around 53 BCM per year throughout the 1990s. The projections point to a small increase in natural gas penetration in all sectors, and very slow growth overall.

Demand management

A range of options relating to demand management may be extremely important in the period up to 2000, particularly with the creation of a privatised utility. The policy of BGC over the past fifteen years has been to emphasise the 'premium sectors' - the residential sector and high value industrial users - with sufficient interruptible customers to balance the load. There has been no attempt to compete in the power generation market. One of the difficulties in trying to project future demand is that the shape of the curve is not known with any certainty. Because of relatively conservative pricing policies by BGC, the effect of, on the one hand, sharp price increases in the residential sector relative to other fuels, or an attempt to compete aggressively with other fuels in the industrial and power generation markets, is not easy to predict.

With BGC remaining essentially intact in its transition to the private sector, its power to control the parameters of the domestic market is reinforced and perhaps even enhanced. While other suppliers may compete at the margin, BGC is likely to continue to control the vast majority of gas sold in the UK for the foreseeable future.

As such, within the limits of price competition with other fuels, the utility will have considerable freedom to decide where - both sectorally and geographically - it wishes to compete, and probably much more important, where it does not. Sectorally, there is little doubt that gas will retain its residential market and may continue to make further slow progress. It is in the industrial sector (where it is intended that prices will be set without any government or regulatory interference) that the main choices remain to be made. Skea has drawn attention to the high degree of price sensitivity in the interruptible gas market. His analysis and projections up to 2010 suggest potential for variations in demand in the order of 2-3 BCM per year, depending on competition with other fuels, principally coal.[98]

Whereas the argument in the past has been that the Corporation needed the industrial sector to balance the load swings caused by the large domestic market, infrastructural developments over the past decade have weakened this imperative. Taken together, the ability of the very large Morecambe field to deliver 1.2 bcf/d, additional offshore storage in the Southern Basin, onshore LNG storage capacity, and arrangements with producers to increase production over short periods constitute a formidable array of load-balancing facilities which, even if not eliminating the need for interruptible customers, have considerably reduced their importance within the total picture.

In addition to the ability to supply peak shaving gas, there are geographical factors which may be of considerable importance in future demand strategy. Catherine Price has indicated the significance to the end user that both peak load pricing and regional pricing for transmission to different geographical areas could assume, if the utility were to adopt a different tariff system.[99] In any event, a privatised utility will need to be somewhat more aware of these cost elements in its capacity as a common carrier being required to convey gas from suppliers to consumers. These charges will be under the supervision of the regulatory authority and the utility will be required to separate out - or in American parlance 'unbundle' - its different services and place a value on each of them. In so doing, the profitability of serving individual sectors and geographical regions should become apparent, and any cross-subsidisation may need to be justified within the Corporation, if not to the outside world.

Thus around the central scenario of static or slightly rising gas demand up to the end of the century, there are other possibilities depending on the future policy of the utility. However, before outlining these possibilities, there is a further aspect of demand management which must be considered.

Efficiency and environment

With long overdue heightening of consciousness in the UK about energy savings and efficiency, one aspect of demand management could be more aggressive promotion of conservation measures in the natural gas sector. In testimony to the Select Committee, BGC foresaw that its measures to promote conservation would reduce demand by the year 2000 by 15%.[100] Conservationists have continually questioned whether BGC (and indeed all of the energy industries) have exerted sufficient efforts to promote savings and efficiency measures among end-users in this sector. In particular, the Association for the Conservation of Energy (ACE) has pointed to the situation in the United States where utilities provide free energy audits for customers and are prepared to extend loans to fund energy conservation measures.[101]

The evidence which ACE has produced requires serious consideration, although it is rather more persuasive (at least to this author) in the electricity than in the gas sector. The situation for many US utilities is that they are so greatly restricted by their regulatory authorities, in respect of the infrastructure which they are permitted to build in order to obtain new supplies, that energy conservation measures become the most cost-effective. It is perhaps not a coincidence that ACE is in favour of a strict regulatory regime for the UK.[102]

None of this is to deny the basic argument of the conservationists, reiterated by the Select Committee in their report: that a careful assessment should be made between investments in incremental sources of supply and in measures which could be taken to create a comparable effect by restricting demand.[103] Despite the view of BGC's Chairman that this was a matter for Government rather than the gas industry, there are future circumstances in which measures to restrict demand rather than promote additional supply could certainly be more attractive to BGC.[104]

If greater efficiency measures work towards restricting gas demand, increasing emphasis on environmental protection could create presssures in the opposite direction. British Governments have been slower than many of their European counterparts to respond to demands for stricter control of industrial emissions. However, this situation may change in the future, as a result of either domestic political pressures or European Community legislation. Imposition of stricter emission controls on coal-burning power stations and industrial boilers might improve the competitiveness of natural gas, particularly in the industrial sector.

Energy demand - high and low scenarios.

There are two scenarios which can be put forward, differing considerably from the 'central case' of annual gas demand of around 55 BCM in the period 1990-2000, and following the extremes of the projections in Table 16.

The first would be a high scenario with demand reaching 65 BCM by 2000. A figure of this magnitude has been most often advanced by Peter Odell along the following lines: gas reserves in the North Sea are much larger than any company or Government has been prepared to admit. These can be produced very cheaply and if they cannot then taxation should be lowered in order to ensure that they can. Gas prices would then be sufficiently low to allow gas to penetrate into lower value industrial uses and power generation. It is only collusion between producers, Government and BGC which prevented this situation occurring in the late 1960s and 1970s.[105]

The low scenario with demand around 40 BCM by 2000 could be brought about by the following situation. Economic growth and industrial production in the UK remain weak and, as a result, energy demand remains stagnant. Following privatisation, the utility's analysis of profitability per therm sold indicates that prices to domestic consumers should be raised to the maximum possible extent and interruptibles should be phased out. Installed storage capacity allows such a policy to be followed with minimal inconvenience. Negotiations between BGC and the producers continue to stall on the issue of prices. Government remains uncooperative on the question of imported gas. The utility - unable to guarantee supplies at what it believes to be competitive prices and with an obligation to provide high returns for shareholders - decides to move towards a smaller, more profitable gas market. Strong promotion of energy efficiency measures assists in further restricting demand, and wins the approval of both the Government and the regulatory agency.

At present, neither of these scenarios seems to this author more likely than the central case, although the low scenario appears more convincing than any move in the opposite direction. In any event, the time horizon of the end of the century is too short to be really interesting for strategic planning. Thus the claim that the UK can be self-sufficient in gas up to the end of the century seems to assume that there is no future beyond 2000. While this is understandable in terms of a reluctance to put forward meaningful figures beyond a fifteen-year time frame, it does not allow a sufficiently detailed examination of the consequences of following a particular policy in the short to medium term.

Supply, Demand and Trade Options to 2020

Looking at the UK gas sector up to the end of the century, there is a range of future supply and demand options which could be selected. Some of those options would place considerable constraints on the sector in the first two decades of the next century. It is therefore important to bear in mind a time horizon longer than the end of the century, otherwise the consequences of 10-15 year policy measures are not sufficiently clear.

The basic considerations for the future are: the size of the gas market; the division of supply between indigenous production and imports; and the rate at which UKCS gas should be produced.

These considerations are addressed below, with the following assumptions: the size of the resource base is as shown in the 1985 Brown Book (Table 2); the situation in 1990 is held constant and assumed to be already determined, with roughly 45 BCM of production and 10 BCM of imports; and the contract for Norwegian Frigg imports expires in 1992/3.

	1. No Imports	2. Imports Allowed		Demand
	UKCS Prod.	UKCS Prod.	Imports	
1990	45*	45	10	55
2000	55	35	20	55
2010	55	35	20	55
2020	55	35	20	55

* in 1990 there will still be about 10 BCM of contracted imports from Norway.

The table above shows two very simple scenarios for the UK gas sector in the period 1990-2020 with a stable demand of 55 BCM - slightly higher than the 1984 level - throughout the entire period. In Case 1, the country imports no gas after the present Norwegian contract runs out in the early 1990s. In Case 2, the country increases its dependence upon imports during the 1990s and continues the same level of imports up to 2020.

In Case 1, all proven reserves and 15 BCM of probable reserves have been produced by the end of the century; by 2020 only 220 BCM of possible reserves remain on the UKCS.[106] In Case 2, about 75 BCM of proven reserves remain to be produced at the end of the century; by 2020 almost all of the proven and probable resource base has been produced, leaving more than 640 BCM of possible reserves to be produced.

	3. High Price/Low Demand			4. Low Price/High Demand		
	UKCS Prod.	Imports	Demand	UKCS Prod.	Imports	Demand
1990	45	10	55	45	10	55
2000	40	0	40	50	15	65
2010	35	0	35	45	20	65
2020	30	0	30	40	25	65

Cases 3 and 4 introduce additional features of the market. In Case 3, higher prices (i.e. higher prices paid to producers which lead to higher prices paid by consumers) restrict demand, to the point where the country is entirely supplied from the UKCS without imports. In Case 4 lower prices allow demand to rise steadily; UKCS production peaks at the end of the century and falls; imports rise in order to fill the gap. The implications for reserves are similar to the previous examples: in Case 3, 55 BCM of proven reserves remain to be produced at the end of the century; 55 BCM of probable reserves and all 640 BCM of possible reserves remain to be produced in 2020. In Case 4, proven reserves are virtually eliminated by 2000; 400 BCM of possible reserves remain after 2020.

The assumptions underlying these two scenarios are as follows. In Case 3, imports are prohibited after the expiry of the Frigg contract and negotiations between the utility and the producers lead to smaller quantities being produced at higher prices, the argument of the purchaser being that markets cannot be retained at high price levels. In Case 4, prices are sufficiently low to encourage substantial market expansion, but in order for this to happen, lower priced imports are 'rolled in' with UKCS production. In this scenario, imports prior to 2000 are very competitive with UKCS gas, and after 2000 the utility may import gas at a lower price than will be paid for new domestic gas developments. This position is highly dependent upon imports of Soviet gas after 2000.

5. Author's Scenario

	UKCS Production	Imports	Demand
1990	45	10	55
2000	45	10	55
2010	40	15	55
2020	35	20	55

In Case 5, I have tried to construct what would seem a reasonable balance between indigenous production and imports, within the context of the resource base, keeping roughly the same sized UK gas market which exists in the mid-1980s. This production profile leaves 35 BCM of proven reserves to be produced after 2000, and more than 500 BCM of possible reserves still exist for the period after 2020. UKCS producers retain the lion's share of the market; and by 2020, the country is dependent upon imports for only just over one third of demand.

In this scenario, imports are seen to be competitive with UKCS production throughout the entire period, particularly after 2000 when costs (and hence prices offered for indigenous gas) may rise sharply. If imports are not available at competitive prices or if this course of action is politically unacceptable, the market is likely to contract as in Case 3. If, however, UKCS costs do not rise and/or additional volumes of gas are discovered which can be competitively produced and imported, the market could expand as shown in Case 4.

Case 5, attempts to be fair to all parties:

The consumer and the nation - by maintaining the market at its present size; depleting the resource base at a brisk, but not irresponsible, rate over the next several decades, leaving reserves to production ratio of more than 14 years in 2020; and ensuring that there is a gradual transition to greater import dependence from 20-25% in 1990, to 35-40% in 2020.
Producers - by maintaining a high level of production through the first decade of the next century.
The utility - by allowing the choice between UKCS production and imports, given a large and stable market size.

These rather artificial scenarios make some simple points about the consequences of different rates of depletion of UKCS gas and the role of imports. They bring out the point that the rejection of the Sleipner import by the Government was a landmark decision for the UK gas sector which has set a course for the nation's demand to be met from its own resources up to the end of the century. If the UK gas market maintains its present size up to 2020, and imports of gas are not permitted, all UK proven reserves will have been produced by the end of the century, and very little of the resource base remains by 2020. If imports continue to be disallowed by the Government, the utility is most likely to deal with this situation by pressing for higher consumer prices and maximum energy efficiency measures in order to reduce the size of the market.

UKCS production up to the end of the century will consist mainly of dry gas with some associated gas and gas from condensate fields. In the period 2000-2020 the situation is reversed, with production coming predominantly from condensate fields, with some associated and dry gas. Because of the higher production and transmission costs for associated gas and condensate fields, it is likely that higher prices will need to be paid in order to encourage large-scale production of these resources. Uncertainties over costs of production and transportation, and the fiscal regime, make it extremely difficult to gauge just how much more expensive these sources will be, compared with present dry gas developments. However, to the extent that high prices are paid to UKCS producers for gas from condensate fields, pressure will be generated to raise consumer prices sharply in real terms around the end of the century.

<u>Trade</u>.   The role of imports within these scenarios needs to be elaborated in slightly more detail. In Case 5, I am suggesting that imports remain competitive with UKCS gas throughout the period under consideration; imports might even be cheaper after the end of the century, depending on UKCS costs of production.   The volumes of 10-15 BCM up to 2010 can be easily supplied by Norway (Sleipner or equivalent) and, if available, a small quantity from the Netherlands.   The increase in imports to 20 BCM per year by 2020 could probably be covered totally by Norwegian imports, but there must be a questionmark as to whether, for security reasons, any one source should be allowed to supply more than 30% of the country's total gas requirements. Because of this, and entering into the spirit of making the UK gas sector a more competitive environment, additional deliveries should probably come from the USSR.   It would be desirable for the first tranche of imports in the period 1993-2010 to be allocated on a competitive basis, but it is not likely that the USSR would be considered as a supplier, because of political and strategic unacceptability, unless this can be negotiated as a displacement agreement with the Netherlands, which would require a large diameter pipeline link (as opposed to a small peak sharing arrangement) with the Continent. It is only if the higher demand scenario of Case 4 becomes relevant that the USSR would be supplying the UK with about 10 BCM per year prior to 2010, probably through a dedicated pipeline.

Thus, for the next two decades, UK gas supplies will almost certainly consist of indigenous production combined with pipeline gas imports from Norway.   Additional supplies could be sought from the Netherlands, if available, and the USSR, if politically acceptable. Pipeline imports from further afield - Africa and the Middle East - and/or large-scale LNG imports seem improbable until at least the second decade of the next century. All indigenous and imported supply options are commercially more attractive than large-scale production of SNG.

There will need to be a thorough evaluation of the possibility of supplies from the USSR. At the present, it is unlikely that Government would be prepared to countenance a base load Soviet gas import. If Soviet gas should continue to be excluded for political/strategic, as opposed to commercial, reasons, this would eliminate an important supply option for the gas sector, which must be a strong contender for consideration in the mid to late 1990s. Supplies from the USSR either directly, or by displacement via the Netherlands, would give rise to a link with the Continental gas grid which might improve European gas security as a whole. Security arrangements might include provision for deliveries of UK gas to Continental countries in an emergency.

This is likely to be the only context in which exports will be seriously considered. There was a time when co-operation with first the Dutch, and then the Norwegians, might have been a good idea, within the context of base-load import contracts; however, these opportunities were missed and unless new reserves are discovered close to median lines, are unlikely to be resuscitated. A large-scale pipeline link to the Continent will only be constructed in the context of a base load import and may, therefore, be delayed until well into the 1990's or even the next century.

In general, there is not likely to be any substantial 'free market' development in the UK natural gas trade. Irrespective of the official policy in this area, Government is certain to retain total control of natural gas trade, making judgements on a case-by-case basis. It may do this through the regulatory office, or directly by means of the Secretary of State issuing licenses for imports and reserving the right to withhold permission in the national interest. There is no country in the world where Government does not have the right to the final decision regarding natural gas imports (and exports) and no reason why the UK should be an exception in this regard. Balance-of-payments considerations will always be a factor in the calculation of Governments regarding the need and timing of imports. In times of financial stringency, these considerations may carry more weight than questions of resource depletion and energy policy.

Depletion policy. What is still at issue is the rate at which the domestic resource base should be depleted and the mix of supplies at any point in time between domestic production and imports. It is in this area that the public debate, including the Select Committee Reports, has been most disappointing. In the mid-1980s, the fashionable view is that such matters should be decided by 'the market', despite the fact that the scope for any kind of gas-against-gas competition has always been extremely limited and will remain so (irrespective of the form that any privatisation of BGC might take). There is considerable emphasis by Government on the need to deplete the resource base at the fastest possible rate. The value of natural gas in the ground is thought to be less than gas produced, and therefore any notion of deferring a proportion of UK gas production to some future date has become undesirable. Imports which would hold back a proportion of domestic gas production are not regarded as beneficial, principally because this would reduce government revenues from the UKCS as well as contributing adversely to the balance of payments.

The problem with such a policy is that rapid, early depletion of lower cost UK reserves will lead either to increased dependence on imported supplies by the late 1990s and certainly in the early part of the next century, or to a smaller UK gas market over the same period. Only those who are certain that a much larger resource base exists than is currently shown in the Department of Energy Brown Book statistics can argue with confidence against these conclusions. Case 5, which I have advanced as the most desirable outcome, would only leave 35 BCM of presently known proven reserves to be produced at the end of the century, and it could be argued that, from a national point of view, even this may be too fast a rate of depletion. However, producers must retain an incentive to explore for new gas and it is highly undesirable that we return to the situation in the 1970s when this was manifestly not so.

I have suggested that, in the period up to 2020, it is possible to provide the nation with the present level of gas supplies while at the same time maintaining incentives for producers. Up to the end of the century, there is the option to achieve this by rapid development of dry gas reserves and excluding imports. If, however, imports are excluded and unless considerable quantities of additional dry gas reserves are located which can be produced at similar cost to present fields, the most likely outcome of pursuing this option will be either a much larger dependence upon imports after 2000, or a much smaller gas market. Present trends and policies would appear to be leading in the latter direction. Thus repeated claims that the country could be self-sufficient in gas up to the end of the century fail to draw attention to the question as to whether this would be an appropriate longer-term policy for the gas sector and the energy balance, or simply a short- to medium-term strategy for maximising revenues.

What is proposed here is not an all-embracing, inflexible strategy for the gas sector over the next several decades; it leaves room for changes of direction in order to accommodate changing availability of supplies (both domestic and imported), and changes in demand. The main hope must be that the option to deplete the UK resource base at the maximum possible rate will not be adopted simply because it may be the easiest course of action from a short- to medium-term financial point of view. The worst possible outcome for the gas sector would be a very rapid rate of production over the next 15-20 years followed by the development of major dependence on imports, or a dramatic reduction in the size of the market. This would be particularly unfortunate given the opportunities available in the late 1980s to arrange a mix of indigenous production and imports from secure sources, which would ensure that the natural gas sector need not become an area of vulnerability in the nation's energy balance over the next three decades.

Notes

[1] Trevor I. Williams, A History of the British Gas Industry. Oxford University Press, 1981, p.296.

[2] Since the early 1970s the Department of Energy has produced a yearly publication entitled Development of the Oil and Gas Resources of the United Kingdom, colloquially known as the 'Brown Book'. It will be so designated in this study with the appropiate year.

[3] Indeed, there have been suggestions that Frigg could be declining by the late 1980s. See the testimony of Sir Denis Rooke in: House of Commons Energy Committee Seventh Report, Session 1984-85, HC 76-I and HC 76-II, The Development and Depletion of the United Kingdom's Gas Resources, HMSO, 19 and 24 July, 1985. (Henceforth, HC 76-I, HC-76 II) p. 281.

[4] Ray Dafter, 'Jobs Boost from Morecambe Field', Financial Times, special on The North West, p.V.

[5] International Energy Agency, Natural Gas: Prospects to 2000. Paris: IEA/OECD, 1982, p.69.

[6] Martin Lovegrove, Lovegrove's Guide to Britain's North Sea Oil and Gas. Cambridge: Energy Publications, 2nd Edition, 1983, p.103.

[7] Department of Energy, Gas Gathering Pipeline Systems in the North Sea, Energy Paper No. 30. London: HMSO, May 1978.

[8] Department of Energy, A North Sea Gas Gathering System, Energy Paper No. 44. London: HMSO, June 1980; also, W. J. Walters, R. H. Wilmott, I. J. Hartill, 'The Northern North Sea Gas Gathering System', Institution of Gas Engineers, Communication 1149, 1981.

[9] Sue Cameron and Ray Dafter, 'North Sea Gas Pipeline Plans Changed', Financial Times, 4 April 1981.

[10] 'Gas Gathering Pipeline', Department of Energy Press Notice, Reference No. 152, 11 September 1981; Ray Dafter, 'Anatomy of a 2.7bn Decision', Financial Times, 1 September 1981.

[11] 'UK Statfjord Gas to be Piped Ashore Via FLAGS', Lloyds List, 17 January 1983; Maurice Samuelson, 'First Statfjord Delivery Taken by British Gas,' Financial Times, 9 October, 1985.

[12]   'Shell unveils the route for the Fulmar line', World Gas Report, 29 August 1983.

[13]   Brown Book, 1985, Appendix 11, pp. 71-2.

[14]   A North Sea Gas Gathering System, op.cit., pp.26-32.

[15]   Dominic Lawson, 'Oil Industry Rejects Plan for Gas Gathering Pipeline', Financial Times, 10 January 1984.

[16]   C. H. Bayly and T. F. Cox, The Economics and Politics of Gas/Condensate Gathering in the UK North Sea. Byfleet: Gaffney Cline and Associates, 1983.

[17]   House of Commons Select Committee on Energy Third Report, Session 1981-82, North Sea Oil Depletion Policy. London: HMSO, 7 May 1982, and North Sea Oil Depletion Policy: The Government's Observations on the Committee's Third Report of Session 1981-82, Sessions 1982-83. London: HMSO, 21 December 1982, pp.x-xi.

[18]   This does not include associated gas used on production platforms (see Table 4).

[19]   HC 76-I, p. xxxv.

[20]   HC 76-II, p. 144.

[21]   Morton Frisch 'The Supply Outlook for Gas from the North Sea Basin', a paper presented to the Seminar, The Gas Supply Outlook for Europe to the Year 2000, Institute of Petroleum, London, 20 May 1985.

[22]   'The Background to the Far North Liquids and Associated Gas Systems', Shell Press Release, May 1982.

[23]   As note [3].

[24]   UK Offshore Operators Association, Potential Oil and Gas Production from the UK Offshore to the Year 2000 UKOOA, September 1984, p. 36.

[25]   Calculated using Brown Book, 1983, Table 8, p.22 and subtracting Algerian imports.

[26]   See note [3].

[27]   House of Commons Energy Committee, Eighth Report, Session 1983-84, HC-438, The British Gas Corporation's Proposed Purchase of Gas from the Sleipner Field HMSO, 21 May 1984. (Henceforth, HC-438.)

[28]   The remainder of this paragraph is taken from Jonathan P. Stern, 'After Sleipner: A Policy for UK Gas Supplies,' Energy Policy, February 1985.

[29]   The details and statistics underpinning this view can be found in Jonathan P. Stern, International Gas Trade in Europe: The Policies of Exporting and Importing Countries, Energy Paper No.8. Aldershot: Gower, 1984, pp. 4-41.

[30]   See for example the presentation of the President of Statoil to the Third European Gas Conference, Oslo, Norwegian Petroleum Society, 25-26 September 1985: Arve Johnsen, <u>Norwegian Natural Gas</u>.

[31]   Klaus Liesen, <u>The West German Natural Gas Industry Between Market Needs and Supply Potentials.</u> Third European Gas Conference, Oslo, Norwegian Petroleum Society, 25-26 September 1985.

[32]   There is the possibility that future Norwegian gas export contracts will not be sold on a field by field basis. Thus it would be simply a question of a buyer signing for a specific quantity of gas of a specific calorific value, at a specific price. The sellers would then make the decision which fields that gas would come from. 'Norway Examines Abandoning Field Dedicated Sales,' <u>International Gas Report</u>, No. 27, 29 March 1985.

[33]   Peter Hinde, <u>Fortune in the North Sea</u>. London: G. T. Foulis & Co, 1966, p.169.

[34]   A.H.P. Grotens, <u>The Dutch Natural Gas Policy</u>, Third European Gas Conference, Oslo, Norwegian Petroleum Society, 25-26 September 1985.

[35]   <u>Ibid</u>.

[36]   Ian Hargreaves, 'British Gas Told to Consider Dutch Deal', <u>Financial Times</u>, 1 March 1984.

[37]   Adrian Hamilton, 'Britain's Options in the Gas Glut', <u>Observer</u>, 15 April 1984.

[38]   At 1 January 1985, Cedigaz estimated Soviet gas reserves at 37.5 TCM 38.9% of world proven reserves. Cedigaz, <u>Le Gaz Naturel Dans Le Monde, en 1984</u> Paris: Cedigaz, June 1985. Table 3, p.10.

[39]   In fact, there is not as much as 10 BCM of spare capacity in the Mittel Europaische Gasleitungsgesellschaft system. However, it is likely that capacity which does exist could be expanded within this time frame, using a variety of different transmission options.

[40]   International Energy Agency <u>Energy Policies and Programmes of the IEA Countries, 1983 Review.</u> Paris:OECD/IEA 1984, Appendix A, Annex 1, pp. 72-3.

[41]   House of Lords Select Committee of the European Communities, Session 1981-82, <u>Natural Gas</u>. London: HMSO, 29 June 1982, p.18.

[42]   Department of Energy, <u>Energy Policy: A Consultative Document</u>, Cmnd 7101. London: HMSO, 1978, p.43.

[43]   'Goodbye Sleipner - For Now Anyway,' <u>International Gas Report</u>, 15 February 1985, p.2.

[44]   See the testimony of Sir Alistair Frame and Allen Sykes in <u>HC 76-II</u>, pp. 312-3; also the testimony of the Confederation of British Industries in House of Commons Energy Committee, Session 1985-86, <u>Regulation of the Gas Industry</u>, HC-15, 15 January 1986. p.118.

[45] Quoted in Dominic Lawson, 'Moscow to Cut French Gas Price', Financial Times, 27 June 1985.

[46] Stern, International Gas Trade, op. cit. pp. 72-104.

[47] Ibid., pp. 125-132, 134-5.

[48] As note [42].

[49] House of Lords, op. cit. p. 18.

[50] UK Department of Energy Press Notice, 10 February 1982. The same view was reiterated by the Minister of State for energy in 1985, HC 76-II, p.304.

[51] For the full text of the landing requirement and BGC's official views on exports see HC 76-II, pp. 26-28.

[52] Colin Robinson, Liberalising the British Gas Market, a paper presented to the Benelux Association of Energy Economists Conference, Luxembourg, 23-25 September 1985. See also Professor Robinson's evidence to the Select Committee, HC 76-II, pp. 209-17.

[53] HC 76-II, pp. 61 and 146.

[54] House of Commons Select Committee on Energy, Third Report, Session 1981-82, North Sea Oil Depletion Policy, op.cit. p. xxii.

[55] In Continental gas trade contracts, no country is entitled to receive actual molecules of gas from any particular source. The contract with the pipeline company simply states volume, quality, and delivery profile; R. Schottker, The Control of the European Natural Gas Transmission System, a paper delivered to the Conference on The Planning and Financing of International Pipelines, London, Scientific Surveys, 14/15 November 1985.

[56] International Energy Agency, Energy Policies and Programmes in IEA Countries. Paris: OECD/IEA, 1983 and 1984, pp. 454 and 496 respectively.

[57] The Mittel Europaische Gasleitungsgesellschaft (MEGAL) pipeline system brings gas from east to west in Europe, and the Trans-Europa Naturgas (TENP) pipeline system brings gas from the North Sea to central and southern Europe. Schottker, loc.cit, contains details of the enormous complexity of the system which allows gas to be shared out between the Continental countries.

[58] 'Algeria's LNG Deal with UK Avoids Price Index Issue', Petroleum Intelligence Weekly, 5 January 1981. Neil Sinclair, 'British Gas May Rejoin Spot Market for LNG', Lloyds List, 27 July 1983.

[59] House of Lords, op.cit., p.103.

[60] Ibid.

[61]   Health and Safety Executive, Canvey: An Investigation of Potential Hazards from Operations in the Canvey Island/Thurrock Area. London: HMSO, 1978. Canvey: A Second Report. London: HMSO, 1981. Department of the Environment, The British Gas Methane Terminal on Canvey Island, Department of Environment, 1983.

[62]   For details of Nigeria see Stern, International Gas Trade, op.cit., pp. 125-31. For information on Qatar and Middle East gas in general see Jonathan P. Stern, Natural Gas Trade in North America and Asia. Aldershot: Gower, 1985, pp. 187-97.

[63]   'British Gas Demonstrate Extended Gas Making Run at Westfield in Scotland', British Gas Press Information, 12 December 1981.

[64]   Commission on Energy and the Environment, Coal and the Environment. London: HMSO, 1981, p.123; $50 per barrel is a Shell estimate from House of Lords, op.cit. p.32.

[65]   Coal and the Environment, op.cit, p.130.

[66]   Ibid., p.131.

[67]   House of Lords, op.cit, p.104.

[68]   'The £20 Billion Gasbags', The Economist, 16 April 1983, p.47.

[69]   Apart from one very small contract in which 30,000 cubic feet per day of supplies from an onshore field will be supplied direct to an industrial user. 'First UK private deal gets green light,' World Gas Report, 13 January 1986, p. 8.

[70]   Oil company arguments in HC-76, II.

[71]   Evidence given by BGC to the House of Commons Energy Committee, First Report of the Energy Committee, Session 1983-84, Electricity and Gas Prices, HC 276-I and HC 276-II, p. 27.

[72]   Stern, 'After Sleipner', loc. cit.

[73]   The figure of 12 pence and the Frigg price of 28 pence in the next paragraph are calculated using the Digest of UK Energy Statistics 1985, Tables 46 and 72, pp. 69 and 101, and the BGC Annual Report, Appendix II, pp. 50-51. These are not quite compatible because the Digest statistics are calendar year and BGC figures are financial year. My intuitive guess is that the domestic figure is too low (and should be nearer 15-16 pence) and the Frigg figure too high, but it should be remembered that sterling was at or near parity with the dollar for much of 1984.

[74]   See previous note.

[75]   It remains to be seen whether the price which is agreed upon will actually remain in force for the long period (perhaps eight years) between signing the contract and first delivery of the gas. Still fresh in our minds is the Statfjord contract, signed in 1980 with a base price of $5.50 per mmbtu, where the price was still around

$4.50 per mmbtu by the time of the first delivery in late 1985. The buyers have managed to reduce the price of deliveries in the initial period of the contract and it is not certain whether the base price and escalation clauses in the original contract will continue to have relevance in determining prices.

[76]   The issue of currency denomination in Soviet gas export contracts is complicated and much depends on the bilateral trade profile between the USSR and the importer. Most European countries pay for their Soviet gas in their own currency, although some contracts are denominated in dollars. In the 1990s, gas imports paid in sterling might prove to be a rather attractive proposition, particularly if they involved a reciprocal export of large quantities of steel pipe and engineering goods.

[77]   'Algeria's LNG Deal with UK Avoids Price Issue', Petroleum Intelligence Weekly, 5 January 1981.

[78]   Except for one spot cargo during the winter. The price established for the nine-month extension - equivalent to some 29-30 pence per therm f.o.b. - was for a small volume of 0.75 BCM for a short and limited period. In other words, these could be regarded as limited peak load, as opposed to (Norwegian) base load, supplies being delivered through facilities where the capital cost had already been amortised over the life of the contract.

[79]   House of Lords, op.cit., p.12.

[80]   Algerian pricing policy and the history of the Continental contracts are outlined in Stern, International Gas Trade, op.cit., Chapter 4.

[81]   Hinde, op.cit., p.109.

[82]   BGC, Annual Report and Accounts, 1982/83, pp.10-13. When questioned on this subject by the Energy Committee with respect to the duties of the new Director General of the Office of Gas Supplies, the Director General of the Office of Telecommunications stated 'There is a social obligation to provide universal services at fixed prices. There is a social obligation to provide services in rural areas. There are social obligations for disabled people, people who are deaf and people who are blind . . . What there is no provision for is to subsidise the bills of people on relatively low incomes. One might take the view, I think, that that was more a matter for the Social Services Department than Trade and Industry.' op.cit., HC-15, 15 January 1986, p. 56.

[83]   John Vickers and George Yarrow, Privatisation and the Natural Monopolies. London: Public Policy Centre, 1985, pp. 10-11.

[84]   Deloitte, Haskins and Sells, British Gas Efficiency Study for the British Gas Corporation and the Department of Energy, June 1983, p.50.

[85]   Vickers and Yarrow, op.cit. pp. 7-10.

[86]   David M. Newbery, 'Pricing Policy,' in Robert·Belgrave and Margaret Cornell (eds), Energy Self-Sufficiency for the UK?, Aldershot: Gower, 1985, pp. 77-120.

[87] HC 276-II, pp. 26-28.

[88] HC 276-I, p. xxxvii.

[89] E.M. Hammond, D.R. Helm and D.J. Thompson, 'British Gas: Options for Privatisation,' Fiscal Studies, Vol. 6, No. 4, November 1985, pp. 1-19.

[90] Deloitte, Haskins and Sells, op.cit., p.36.

[91] House of Commons Public Accounts Committee, Fifth Report, Session 1985-86, Monitoring and Control of Investment by the Nationalised Industries in Fixed Assets, 11 November 1985, p. 5.

[92] Deloitte, Haskins and Sells, op.cit., p.50.

[93] HC-15, pp. xv-xvi.

[94] Department of Energy, Proposed Authorisation to be granted by the Secretary of State for Energy to the British Gas Corporation under Section 7 of the Gas Bill. p.4, Condition 3, Section 3, states:

   (1) If in respect of any Relevant Year the Average Price per therm exceeds the Maximum Average Price per therm by more than 4% of the latter, the Supplier shall furnish an explanation to the Director and in the next following Relevant Year the Supplier shall not effect any increase in prices unless it has demonstrated to the reasonable satisfaction of the Director that the Average Price per therm would not be likely to exceed the Maximum Average Price per therm in that next following Relevant Year.
   (2) If, in respect of any two successive Relevant Years, the sum of the amounts by which the Average Price per therm has exceeded the Maximum Average Price per therm for the second of those years, then in the next following Relevant Year the Supplier shall, if required by the Director, adjust its prices such that the Average Price per therm would not be likely, in the judgement of the Director, to exceed the Maximum Average Price per therm in that next following Relevant Year.
   (3) If in respect of each of two successive Relevant Years the Average Price per therm is less than 90% of the Maximum Average Price per therm, the Director, after consultation with the Supplier, may direct that, in calculating Kt in respect of the next following Relevant Year, there shall be substituted for Tt-1 in the formula set out in paragraph 1 above such figure as the Director may specify being not less than Tt-1 and not more than 0.90 (Qt-1Mt-1)

[95] The best example here is Canada, where the criteria for determining domestic and export prices (and volumes) were for decades (and to a lesser extent still are) subject to a battery of tremendously detailed 'tests' and formulae. Jonathan P. Stern, Natural Gas Trade in North America and Asia, Aldershot: Gower, 1985, pp. 33-44.

[96] Deloitte, Haskins and Sells, op. cit., p.53.

[97] There is, however, little likelihood that the situation will change substantially for the better since all the parties involved will

strenuously resist any attempt to publish terms of contracts, claiming commercial confidentiality.

[98]   HC 76-II, pp. 318-23.

[99]   HC 76-II, pp. 200-4; see also C.M. Price, Distribution Costs in the U.K. Gas Industry, University of Leicester, Department of Economics, Discussion Paper No. 31., January 1984.

[100]   HC 76-II, p.95.

[101]   HC-15, pp. 142-6.

[102]   Ibid., pp. 58-62.

[103]   HC 76-I, pp. xxiv-xxv.

[104]   HC 76-II, pp. 290-1.

[105]   See inter alia, Ibid., pp. 324-38; House of Commons Energy Committee, Eighth Report, Session 1983-84, The British Gas Corporation's Proposed Purchase of Gas from the Sleipner Field, 21 May 1984, pp. 40-44.

[106]   In each of these scenarios, it is assumed that the Morecambe field is used for base load purposes. If the field continues to be used simply for peaking, as at present, this would shorten the baseload lifetime of UKCS reserves.

Appendix: A Note on Northern Ireland

Before summing up the UK situation as a whole and outlining the options for the future, it is worth noting that there is a part of the UK in which natural gas seems unlikely to play any future role, after a number of different possibilities have been discussed involving indigenous and imported supplies.

Northern Ireland's small town-gas industry, composed of thirteen undertakings, was largely dependent upon hydrocarbon feedstocks (naphtha and liquefied petroleum gas) and became uncompetitive after the first oil price rise of 1973/74.[1] In 1976, BGC carried out a study which confirmed the uncompetitive position of the industry and examined a number of alternatives: piped gas from Scotland, piped gas from Eire, importation of LNG and importation of LPG. The option which received most discussion and comment was a pipeline from Scotland which would have brought North Sea gas to Northern Ireland. The BGC report concluded that a link to Northern Ireland would be unprofitable, and would result in losses amounting to £87.5m. by 1989. As a result of this finding, the Government announced, in mid-1979, that pipeline gas to Northern Ireland could not be justified, nor could the continued subsidisation of the industry, which would be assisted to run down in an orderly fashion.

A number of organisations in the Province took issue with the BGC conclusion and commissioned their own report, which suggested a more favourable economic climate for piped gas supplies.[2] The view of one group, committed to maintaining employment in the industry in Northern Ireland, was as follows:[3]

> It is notable that other areas of the UK more remote than Northern Ireland have received supplies of natural gas as of right. Discrimination against Northern Ireland in this matter is scandalous.

(In fact, BGC has no obligation to supply gas 'as of right' and there are many areas of the UK which do not have a piped gas supply.)

The conclusion of a careful and well-balanced study of the question was that pipeline gas from Scotland to Northern Ireland could be economic, but that a decision should be made dependent upon negotiations with the Republic of Ireland on the cost and availability of gas supplies from the Kinsale field, which are being piped to Dublin via a line which could be extended north to Belfast.[4] Projected gas demand was estimated at around 80 million cubic metres in the early 1980s, rising to 275 million cubic metres by 1989. These are not large figures and would not have constituted a great drain on UK reserves, had the pipeline from Scotland been built. However, the cost of a pipeline from Scotland to Northern Ireland had been estimated at £93m. as against £23m. for the line from the Dublin terminal to Belfast city gate, plus £2.3m. for compression.[5]

In October 1983, agreement was reached on gas sales to Northern Ireland from the Kinsale field, via a pipeline to be constructed and operating by 1985; the contract would run for 22 years.[6] Despite the claim by the Irish Minister for Industry and Energy that 'no political overtones have entered the discussion', it seems clear that there was a considerable political incentive on the British side to achieve an agreement, both in terms of improving relations with the Dublin Government (and creating long-term links between Northern and Southern Ireland) and of creating employment in Ulster.[7] From the outset, the commercial aspect of the trade seemed shaky, with the issue of the price of the gas being left open, although a figure of 26 pence per therm was mentioned.

Less than a year after the agreement had been reached, and as final signature appeared imminent, the contract was cancelled. The British Government had decided that the price - which, with the fall of sterling and the Irish Punt against the dollar, had risen to more than 30 pence per therm - was too high and that the gas would be uncompetitive with coal.[8] Despite a desperate rearguard effort within the Province to find some alternative solution, a decision was taken to close the industry over a three year period at a cost of £97m. and the loss of 1000 jobs.[9]

The Northern Irish natural gas episode emphasised a number of trends in the UK gas sector which may become more pronounced in the future: unwillingness to supply small, geographically remote regions with indigenous gas at relatively high cost, and unwillingness to import gas supplies, even if this means a smaller gas market - or in this case its elimination. While this may not be the end of the story for natural gas in the Province, particularly if the more bullish forecasts of reserve potential in the Republic are to be believed, it has emphatically closed the door on any possibility of supplies from the UKCS.

Notes

[1]   House of Commons Select Committee on Energy, Third Report, Session 1980-81, The Gas Industry in Northern Ireland. London: HMSO, 30 July 1981. The information in this paragraph is drawn from that report.

[2]   Ibid., Appendix 6, pp.28-30.

[3]   Ibid., p.19.

[4]   M. McGurnaghan and S. Scott, Trade and Cooperation in Electricity and Gas, Understanding and Cooperation in Ireland, Cooperation North, Paper 4. Belfast and Dublin: Cooperation North, July 1981.

[5]   Parliamentary reply from the Secretary of State for Energy, Energy Focus, Vol. 1, No.2, July 1984, p. 185. Figures are from October 1983.

[6]   Brendan Keenan and Maurice Samuelson, 'Dublin and Belfast to be Linked by Gas Pipeline', Financial Times, 11 October 1983.

[7]   Ibid. Also Maurice Samuelson, 'King Coal and the Ulster Connection',
        Financial Times, 30 November 1983.

[8]   Brendan Keenan, 'The end of a £500m. Irish Pipe Dream,' Financial
        Times, 5 September 1984.

[9]   'Ulster gas industry faces £97m rundown', Financial Times , 6 April
        1985.

Table 1   Estimates of UKCS Natural Gas Reserves (BCM)

Organisation	Date of estimate	Description of reserve figure	Reserves
1. UK Department of Energy	31 December 1984	Initial proven Initial probable Possible	1,229 600 643
2. British Gas Corporation	1984	Initial proven and probable Possible plus speculative	1,443 594
3. Shell	1984	Producing and developable Significant Discoveries Undiscovered and Leads	1,174 657 458
4. Phillips Petroleum	1985	Proven and probable	1,557
		Total Maximum Resource Potential	3113-3679
5. British Petroleum	1984	Economically recoverable Future Discoveries	2,179 708

Sources: 1. Department of Energy, Development of the Oil and Gas Resources of the United Kingdom (Brown Book), 1985 Table 3, p.9.
2. Seventh Report from the House of Commons Energy Committee, Session 1984-85, Volume II, HC 76-II, The Development and Depletion of the United Kingdom's Gas Resources, HMSO, July 1985, p. 19. See also British Gas Corporation, Britain's Future Gas Supplies 1985-2000: The British Gas View and a Review of Recent Forecasts, November 1984, pp.5-8.
3. British Gas Corporation, op.cit., p.85.
4. Ibid., pp. 67-68.
5. Ibid., p. 139.

Table 2 Estimates of Recoverable Gas Reserves in Present Discoveries on the
UKCS as at 31 December 1984 (BCM)

	Proven	Probable	Proven+ Probable	Possible
A. Initial recoverable reserves:				
From dry gas fields:				
1. a) In production or under development				
i. Southern Basin	714	25	739	20
ii. UK Frigg and Morecambe	187	37	224	31
	901	62	963	51
b) Other significant discoveries not yet fully appraised				
i. Southern Basin	156	210	365	176
ii. UK Frigg and Morecambe	-	-	-	42
Total dry gas	1057	272	1328	269
From gas condensate fields: (a)				
2. a) In production or under development	40	8	48	14

Table 2 (cont'd 1)

	Proven	Probable	Proven+ Probable	Possible
b) Other significant discoveries not yet fully appraised	-	275	275	320
Total gas from condensate fields	40	283	323	334
3. Associated gas from oil fields:(a)				
a) In production or under development:				
i. Currently delivering gas ashore	99	3	102	3
ii. Expected to be connected	31	14	45	8
Sub-total	130	17	147	11
b) Other significant discoveries not yet fully appraised	3	28	31	28
Total associated gas	133	45	178	40
Total initial reserves in present discoveries	1,229	600	1,829	643

Table 2 (cont'd 2)

	Proven	Probable	Proven+ Probable	Possible

Remaining recoverable reserves:

Cumulative production to end of 1984 (b)

1. Dry gas

    a. Southern Basin — 445
    b. UK Frigg and Morecambe — 44

2. Associated gas from oil fields — 15

Total cumulative production to end of 1982 — 504

Total remaining reserves in present discoveries — Proven 725, Probable 600, Proven+Probable 1325, Possible 643

(a) All in Northern Sector of North Sea (North of 56 degrees N.)
(b) Excludes flared gas and gas used on platforms.

Source: Department of Energy, Development of the Oil and Gas Resources of the United Kingdom (Brown Book), 1985 Table 3, p.9.

Table 3   Gas Production from UKCS (million cubic metres) 1976-84

Southern Basin Fields	Total to end 1978	1979	1980	1981	1982	1983	1984	Total to end 1984	R/P Ratio* (Years)
West Sole	19,687	1,365	1,445	1,455	1,512	1,719	1,899	29,082	10
Leman Bank	127,030	13,831	9,482	13,207	11,675	11,985	9,376	196,586	11
Hewett Area	53,809	6,288	6,568	5,048	4,108	3,851	3,631	83,303	8
Indefatigable	40,606	6,006	6,878	5,613	5,720	4,700	5,590	75,113	9
Victor	-	-	-	-	-	-	472	472	43
Viking Area	32,879	4,397	4,689	3,307	4,381	3,413	3,197	56,263	9
Rough	2,516	1,005	467	99	101	27(f)	(55)g	4,270	6
Frigg(a)	3,521	5,345	6,374	7,057	6,569	6,948	7,781	43,595	6
Piper(b)	4	536	521	520	629	782	739	3,732	
FLAGS(c)	-				2,144	4,297	5,356	11,797	
Other(d)	471	455	866	1,098	1,437	1,807	2,066	8,200	
Total(e)	280,523	39,228	37,290	37,404	38,276	39,529	40,162	512,413	
Imports		8.82	10.78	11.06	10.15	10.50	13.16		
		48.05	48.07	48.46	48.43	50.03	53.32		

57

Table 3 (cont'd)

(a) UK share only (39.18%).
(b) Gas used offshore or delivered to land via the Frigg pipeline system.
(c) Gas delivered to land via the Far North Liquid and Associated Gas (FLAG) System.
(d) Associated gas, mainly methane, produced and used mainly on Northern Basin oil production platforms.
(e) Gross production, ie includes own use for drilling, production and pumping, but excludes gas flared.
(f) Production in respect of a two month period only; field remained available for production for most of 1983 whilst work on converting the field to a seasonal storage facility continued.
(g) Production in respect of a 4 month period only.

* Life of reserves in years, calculated by taking remaining recoverable reserves (initial recoverable reserves, given in Appendix 7, pp. 63-66, minus cumulative production) divided by 1984 production.

Sources: Ibid., Appendix 9, p.69.

58

Table 4.

Natural Gas Reserves in Fields in Production and Under Development
in 1985

Southern Basin Dry Gas Fields	Initial Reserves (BCM)	Peak/Plateau Annual (BCM)	Start-up
a. In Production			
West Sole	48	1.4-1.9	1967
Leman Bank	298	9-13	1968
Hewett Area	112	3-6	1969
Indefatigable	127	5-6	1971
Viking Area	84	3-5	1972
Rough*	10	0.5-1	1975
Victor	20	0.5-2	1984
	699		
b. Under Development			
Esmond)			
Forbes)	16		
Gordon)			
South East Inde	13	**	
North Sean)			
)	13		
South Sean)			
	42		
Total	741		
Other Dry Gas			
Frigg (UK)	90	6-8	1977
Morecambe	119	**	1985
Associated Gas			
Brent	91		1982
Statfjord (UK)	16		1985

* Since 1984 the field has been used for storage purposes.
** Used for seasonal deliveries.

Source: Brown Book, 1985, Appendix 7, pp. 63-66; Martin Lovegrove,
Lovegrove's Guide to Britain's North Sea Oil and Gas, 2nd Edition,
Cambridge: Energy Publications, 1983, Table 30, p. 101.

Table 5.   Probable and Possible Gas Developments on UKCS

PROBABLE*

Field	Type/Location	Reserves (BCM)	Annual Peak/ Plateau (BCM)	Expected*** Start-up
Amethyst	Gas/S.Basin	21	1.35	1989
Audrey	Gas/S.Basin	34	2.00	1988
Valiant	Gas/S. Basin	55	4.00	1988
Cleeton/ Ravenspurn	Gas/S. Basin	28	2.00	1990
Barque/ Clipper	Gas/S.Basin	26	2.50	1990
		164	11.85	
Bruce	Condensate/ E. Shetland	69	5.00	1990
Gannet/ Kittiwake	Associated Gas Central Basin	11	1.05	1991
Miller	Associated Gas Central Basin	10	1.20	1990
Total		254	19.10	

POSSIBLE**

Field	Type/Location	Reserves (BCM)	Annual Peak/ Plateau (BCM)	Expected*** Start-up
41/24 & 25	Gas/S.Basin	14	1.00	1993
43/26	Gas/S. Basin	9	0.75	1990
44/22	Gas/S. Basin	28	1.75	1993
47/13	Gas/S. Basin	9	0.75	1991
48/11	Gas/S. Basin	28	1.90	1992
48/18	Gas/S. Basin	10	1.00	1992
49/5	Gas/S. Basin	9	0.75	1993
53/4	Gas/S. Basin	11	1.00	1990
		118	8.90	
Beryl	Associated Gas Northern Basin	26	2.00	1993
Drake	Gas and Condensate Central Basin	17	2.00	1996
29/2	Gas and Condensate Central Basin	17	1.10	1998
Total		178	14.00	

* significant progress expected within 1½-2 years.
** no action expected within 2 years.
*** reflects author's judgement on purchaser's requirements.

Source: Wood Mackenzie & Co., North Sea Report, No. 149, 27 September, 1985, Tables 3 and 4, pp. 6-7.

Table 6   Gas Flaring at Oilfields and Terminals (million cubic metres) 1978-84

	1978	1979	1980	1981	1982	1983	1984	Total to end 1984
Argyll*	40	53	49	30	75	54	22	454
Auk*	47	35	23	27	27	26	22	329
Beatrice	–	–	–	9	31	15	19	73
Beryl	247	201	138	102	108	129	165	1,978
Brae	–	–	–	–	–	192	441	634
Brent	1,598	3,347	1,264	1,221	1,123	585	276	9,982
Buchan*	–	–	–	59	86	99	58	303
Claymore	120	41	12	14	14	8	21	237
South Cormorant	–	5	151	80	56	60	49	403
North Cormorant	–	–	–	–	36	16	27	39
Deveron	–	–	–	–	–	–	1	1
Duncan*	–	–	–	–	–	4	53	58
Dunlin	34	265	224	181	95	65	56	919
Forties	1,172	1,129	911	798	738	739	650	7,711
Fulmar	–	–	–	–	289	292	121	702
Heather*	19	99	35	83	109	71	63	479
Hutton*	–	–	–	–	–	–	21	21
N.W. Hutton	–	–	–	–	–	193	45	239
Magnus	–	–	–	–	–	166	185	351
Maureen*	–	–	–	–	–	63	285	348
Montrose*	182	200	183	151	117	91	98	1,168
Murchison(a)	–	–	47	381	217	180	49	811
Ninian	3	453	552	485	213	138	136	1,979
Piper	933	578	310	185	135	127	180	2,859
Statfjord(b)	–	9	72	19	32	41	34	207
Tartan	–	–	–	124	86	52	16	277
Thistle	151	189	223	209	218	128	73	1,189
Total offshore	4,546	6,604	4,194	4,095	3,805	3,531	3,166	33,787

Table 6 (cont'd)

	1978	1979	1980	1981	1982	1983	1984	Total to end 1984
Total offshore	4,546	6,604	4,194	4,095	3,805	3,531	3,166	33,787
Wytch Farm	–	2	7	2	1	4	5	22
Flotta Terminal(c)	–	–	44	34	17	28	22	478
Sullom Voe Terminal(d)	–	–	–	87	226	196	172	682
St Fergus	–	–	–	–	25	28	22	101
Total onshore	–	2	51	123	269	256	248	1,282
Total gas flared(e)	4,546	6,606	4,245	4,218	4,074	3,787	3,414	35,069

Notes to Table 6

(a) UK share (83.75%) of wellhead production.
(b) UK share (15.9069%) of wellhead production.
(c) Pipeline terminal serving Claymore, Piper and Tartan fields.
(d) Pipeline terminal serving Brent, North and South Cormorant, Dunlin, Heather, Murchison, Ninian and Thistle fields.
(e) Gaseous hydrocarbons associated with crude oil production containing methane, ethane, propane, butane and condensates.

* indicates that all gas which is not used at the platform is flared.

Sources: UK Department of Energy, Digest of United Kingdom Energy Statistics, 1985. London: HMSO, 1985, Table 29, p.45.

Table 7

Investment Requirements for Condensate Fields *

	A	B	C	D	L
Gas Reserves (BCM)	10.5	23.5	41.9	65.7	91.7
Condensate Reserves (million barrels)	35	80	140	220	310
Field Size (million barrels oil equivalent	100	225	400	625	875
Capital Expenditure (a)	700	830	1130	1240	1490
Annual Operating Expenditure	42	50	68	74	89
Peak Production BCM/year	92	166	400	626	875
Peak Oil Production (thousand barrels per day	15	27	34	54	76
Wells	8	14	20	35	40
Platforms	1	1	1	1	2
First Liquid Sale (b)	7	7	7	7	7
First Gas Sale (b)	7	7	15	15	15

(a) 1983 £ m.
(b) years from discovery

* Costs and production profiles for each type of representative field have been developed. Fields A and B are produced by simple depletion (blowdown), fields C,D and L are produced by recycling followed by blowdown.

Source: UK Offshore Operators Association (UKOOA), <u>Potential Oil and Gas Production from the UK Offshore to the year 2000</u>, September 1984, p.52.

Table 8

Investment Requirements for Southern Basin Dry Gas Fields *

	A	B	C	D	L
Field Size (BCM)	2.8	7.1	15.6	28.3	42.5
Capital Expenditure(a)	111	196	302	399	448
Annual Operating Expenditure	6	15	20	25	27
Peak Production mmcfd	27	68	151	205	205
Wells	5	10	22	29	29
Platforms	1	3	4	5	6
Plateau (years)	6	6	6	8	12
Pipeline (miles) (b)	10	20	20	20	20
Swing factor (%)	167	167	167	167	167

(a) 1983 £ million
(b) to existing facilities

* Costs and production profiles for each type of representative field
have been developed. These reflect typical development of fields of
these sizes located in the Southern North Sea. They are not, however,
fully engineered costs for any particular field.

Source: UK Offshore Operators Association (UKOOA), Potential Oil and
Gas Production from the UK Offshore to the year 2000, September 1984,
p.50.

Table 9

Cost Structure of BGC 1974-84

	1974-5	1975-6	1976-7	1977-8	1978-9	1979-80	1980-1	1981-2	1982-3	1983-4	1984-5
Natural gas and feedstock pence per therm	1.68	1.90	2.06	3.07	4.65	5.77	7.85	10.58	11.60	13.26	16.26
% of total cost	20.0	18.5	16.2	20.7	29.3	31.2	32.9	36.3	36.1	41.1	46.1
Operating costs %	80.0	81.5	83.8	79.3	70.7	68.8	63.8	55.7	54.1	49.4	45.9
Gas levy %							3.3	8.0	9.9	9.5	8.0

Source: British Gas Corporation, Annual Report and Accounts, 1984-85, pp. 24, 28-9.

65

Table 10

Cost/Price Reference Frame for Domestic and Imported Base Load Sources of Supply

	pence per therm
Average cost of BGC natural gas supplies 1984-85	16.26
Estimated average price paid for UKCS gas in 1984	12.08
Estimated price paid for Frigg gas in 1984	27.81
Average reported price for BGC purchases of UKCS gas, 1984/85	23-26
Reported price in Norwegian Sleipner contract (turned down by UK Government) end 1984 (£1 = $1) end 1985 (£1 = $1.5)	35 25
Siberian gas landed in UK (estimated end 1985 Continental European 1982 contract price plus allowance for additional transportation to UK paid in sterling).	25-30
SNG from coal (1982 estimate)	53.9

Source: 1984-85 gas supply cost from Table 9, all other figures are illustrative and estimated by the author from fragmentary published material.

Table 11

Average Revenue from Gas Sold to Domestic and Industrial Customers*

(pence per therm)

	DOMESTIC			INDUSTRIAL	
	Price	Increase from Previous Year %		Price	Increase from Previous Year %
1976/77	16.6	15.3		7.2	38.5
1977/78	18.5	11.4		9.7	34.7
1978/79	18.5	0		11.9	22.7
1979/80	19.5	5.4		14.1	18.5
1980/81	24.2	24.1		18.8	33.3
1981/82	30.4	25.6		21.7	15.4
1982/83	37.9	24.7		23.3	7.4
1983/84	39.7	4.7		24.4	4.7
1984/85	41.2	3.8		26.9	10.2

Real Domestic
Price Increases

June 1979	8%
April 1980	17%
October 1980	10%
April 1981	15%
October 1981	10%
April 1982	12%
October 1982	12%
January 1984	4.3%
February 1985	4.5%

* includes standing charges and different tariff categories

Sources: British Gas Corporation, Annual Report and Accounts, 1984-85, Appendix III,
p.52. House of Commons Energy Committee, First Report of the Energy Committee, Session
1983-84, Electricity and Gas Prices, HC 276-I, Table 1, p. xv.

Table 12   Gas Price Formula for Tariff Customers

'The Supplier shall in setting its prices for tariff customers take all reasonable steps, having particular regard to the interests of those customers, to secure that in each Relevant Year its Average Price per therm shall not exceed the Maximum Average Price per therm calculated in accordance with the following formula:

$$Mt = \left(1 + \frac{RPIt}{100} - X\right) Pt-1 + Yt - Kt$$

where

Mt = Maximum Average Price per therm in relevant year t;

$RPIt$= the percentage change (whether of a positive or negative value) in the Retail Price Index between that published with respect to October in Relevant Year t and the preceding October.

X = (value to be determined)

$$Pt-1 = Pt-2 \left(1 + \frac{RPIt-1 - X}{100}\right)$$

but, in relation to the first Relevant Year, $Pt-1$ (and accordingly, in relation to the second Relevant Year, $Pt-2$) shall have a value of ( ) per therm.

Yt = Allowable Gas Cost per therm in Relevant Year t

Kt =   the corection per therm (whether of a positive of negative value to be made in Relevant Year t (other than the first Relevant Year) which is derived from the following formula :

$$Kt = \frac{Tt-1 - (Qt-1Mt-1)}{Qt} \left(1 + \frac{It}{100}\right)$$

in which

$Tt-1$ = Tariff Revenue from Tariff Quantity in Relevant Year $t-1$.

$Qt-1$ = Tariff quantity in Relevant Year $t-1$

Qt = Tariff quantity in Relevant Year t

$Mt-1$ = Maximum Average price per them in Relevant Year $t-1$

It =   the interest rate in Relevant Year t which is equal to, where Kt (taking no account of It for this purpose) has a positive value, the Specified Rate plus 3%, or where Kt (taking no account of It for this purpose) has a negative value, the Specified Rate.

Source: Department of Energy, Proposed Authorisation to be granted by the Secretary of State for Energy to the British Gas Corporation under Section 7 of the Gas Bill. Condition 3, pp. 7-8.

Table 13   Total UK Primary Fuel Consumption and the Natural Gas Position (a)   %

	1978	1979	1980	1981	1982	1983	1984	1985*
Coal	33.3	34.4	35.1	35.5	33.6	33.7	23.9	32.3
Petroleum	44.4	42.5	39.9	38.1	39.0	37.4	46.2	35.3
Natural gas	18.1	18.9	20.6	21.6	21.8	22.7	23.3	25.0
Nuclear power	3.7	3.7	3.8	4.1	4.9	5.5	5.9	6.8
Hydro power	0.6	0.6	0.6	0.7	0.7	0.7	0.6	0.6
Total (mtoe)	211.9	221.4	202.7	196.2	193.7	194.1	193.8	n/a
Natural gas(b) consumption (mtoe)	38.3	41.8	41.8	42.4	42.2	44.0	45.2	47.9
Natural gas production as a % of consumption	88.5	81.8	76.6	75.7	78.2	77.3	74.1	74.1

(a) includes oil and gas for non-energy use and marine bunkers.
(b) includes land and colliery methane and associated gas produced and used mainly on Northern sector oil production platforms. Excludes gas flared or reinjected.

* Preliminary

Source: Brown Book 1985 adapted from Table 14, p.32.

69

Table 14   Sectoral Consumption of Natural Gas* (%)

	Power stations	Iron and steel	Other industry	Domestic	Public admin. and commercial
1977	3.6	3.3	37.0	45.2	10.9
1978	2.2	2.9	36.1	47.4	11.4
1979	1.4	3.2	33.9	49.6	11.9
1980	0.8	2.7	33.3	50.7	12.4
1981	0.5	2.5	31.7	52.7	12.7
1982	0.5	2.2	31.9	52.3	13.1
1983	0.5	2.0	31.4	52.7	13.4
1984	1.0	2.2	31.7	51.6	13.5

* In 1977 town gas amounted to 0.5% of UK gas supply; by 1984 this had fallen to 0.1%.

Source: UK Department of Energy, Digest of United Kingdom Energy Statistics, 1985. London: HMSO, 1985, Table 46, p.69.

Table 15   Alternative UK Gas Demand Estimates 1990 (billion therms a year)

Sector	(1) 1983 ACTUAL	(2)+ DEn High Price	(3)+ DEn Low Price	(4) BGC	(5) Esso	(6) BP Lower Price	(7) BP Higher Price	(8) Shell High Growth	(9) Shell Low Growth	(10) ESI Scenario C
Industry	3.9	3.3–3.8	3.9–4.1	6.5*	6.5*	5.8*	5.7*	6.0*	5.8*	5.3*
Domestic	8.9	9.7–10.8	10.0–11.1	10.6	10.5	10.6	9.1	9.3	9.2	11.3
Other Commercial/ Public Services	2.3	2.4–2.7	2.3–2.7	3.2	2.8	3.0	2.6	2.9	2.7	2.6
Total Energy Use	15.1	15.6–17.3	16.3–17.8	20.3	19.8	19.4	17.4	18.2	17.7	19.2
Non-Energy Use (Feedstocks)	1.8	1.6	1.6	(A S A B O V E *)				(A S A B O V E *)		
Total Final Demand	16.9	17.2–18.9	17.9–19.4	20.3	19.8	19.4	17.4	18.2	17.7	19.2
BGC own use and losses	1.8	1.4–1.5	1.4–1.5	n/a	0.8	n/a	n/a	1.7	1.7	n/a
Total UK Gas Demand	18.8	18.5–20.3	19.3–20.8	n/a	20.6	n/a	n/a	19.9	19.4	n/a
Total in BCM	51.7	50.9–55.8	53.1–57.2	55.8	56.7	53.4	47.9	54.7	53.4	52.8

(i) + initial evidence from the Department. For subsequent revisions, refer to accompanying notes.

71

Table 16   Alternative UK Gas Demand Estimates 2000 (billion therms a year)

Sector	(1) 1983 ACTUAL	(2)+ DEn High Price	(3)+ DEn Low Price	(4) BGC	(5) Esso	(6) BP Lower Price	(7) BP Higher Price	(8) Shell High Growth	(9) Shell Low Growth	(10) ESI Scenario C
Industry	3.9	2.1-2.8	2.8-3.0	6.5*	6.9*	6.3*	5.7*	5.5*	5.4*	3.3*-8.3*
Domestic	8.9	8.6-10.8	10.9-12.5	10.8	11.2	11.6	9.3	9.9	8.9	10.4-11.7
Other Commercial/ Public Services	2.3	2.1-2.5	2.5-2.9	3.2	3.1	3.2	2.8	3.3	2.4	2.2-3.4
Total Energy Use	15.1	12.8-16.0	16.4-18.4	20.5 (18-22)	21.2	21.1	17.8	18.7	16.7	17.0-21.1
Non-Energy Use (Feedstocks)	1.8	0.5	0.5	(A S	A B O V E *	) (	A S	A B O V E	*	)
Total Final Demand	16.9	13.3-16.5	16.9-18.9	20.5 (18-22)	21.2	21.1	17.8	18.7	16.7	17.0-21.1
BGC own use and losses	1.8	1.2-1.4	1.4-1.5	n/a	0.8	2.2	1.3	1.7	1.7	n/a
Total UK Gas Demand	18.8	14.6-17.9	18.3-20.4	n/a	22.1	23.3	19.1	20.4	18.4	n/a
Total in BCM	51.7	40.2-49.2	50.3-56.1	56.4	60.8	64.8	52.5	56.1	50.6	48.6-58.0

(i) + initial evidence from the Department. For subsequent revisions, refer to accompanying notes.

72

Notes and References for Tables 15 and 16

1.   Department of Energy. The Higher Fossil Fuel Price Case in column 2 is taken from Seventh Report from the Energy Committee, HC 76-II, p.15, and the Lower Fossil Fuel Price Case in column 3 from p.16. These data are based upon Energy Projections, 1982 (EP 82). Note that the heading on Table 2, HC 76-II p.16 is incorrect and should read Lower (not Higher). Subsequently, in March 1985, the Department submitted estimates suggesting that industrial gas sales were likely to be some 1 bn therms in 1990 and 1-2.5 bn in 2000 above those of EP 82, (HC 76-II, p.158 and p.294, question 1). The form in which the Department presented its revised estimates has made it difficult to incorporate them in the summary Table.
2.   BGC. The data in column 4 are taken from HC 76-II, p.22. The BGC's data (and those of some other witnesses) were provided in less disaggregated form than those of the Department. The key difference is that BGC's estimates of sales to industry include non-energy use of feedstocks in the chemical sector (as is the case with data provided by other witnesses). See also HC 76-II, p.271.
3.   Esso. The data in column 5 are taken from HC 76-II, p.46.
4.   BP. The data in columns 6 and 7 are taken from HC 76-II, p.64, supplemented by BP, unpublished evidence.
5.   Shell. The data in columns 8 and 9 are taken from the Table on p.119, HC 76-II.
6.   Electricity Council. The date in column 10 are taken from Electricity Council, HC 76-II, p.187. ESI = Electricity Supply Industry.

Source for Tables 15 and 16: Seventh Report from the House of Commons Energy Committee, Session 1984-85, HC 76-I, The Development and Depletion of the United Kingdom's Gas Resources, HMSO, 19 July 1985. Table 5, p.xxi.

MAP 1

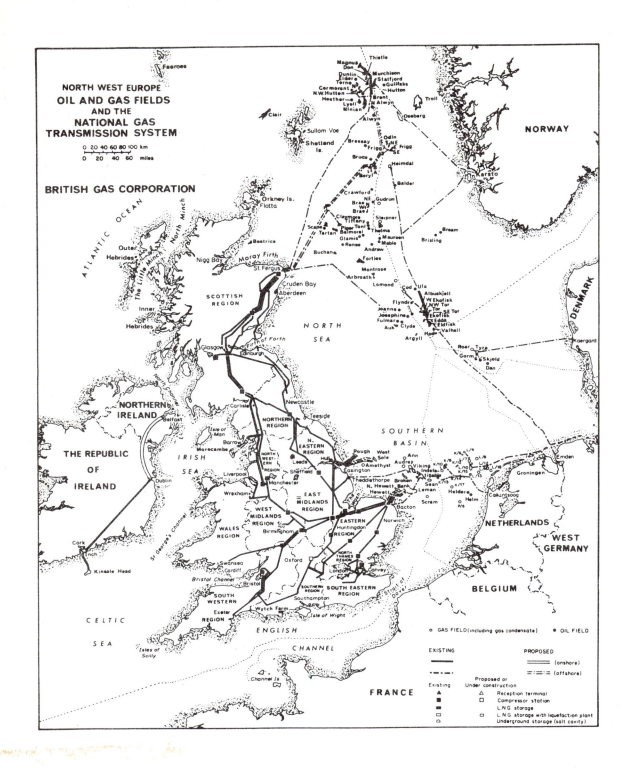

North-West Europe Oil and Gas Fields and UK National Gas Transmission

MAP 2

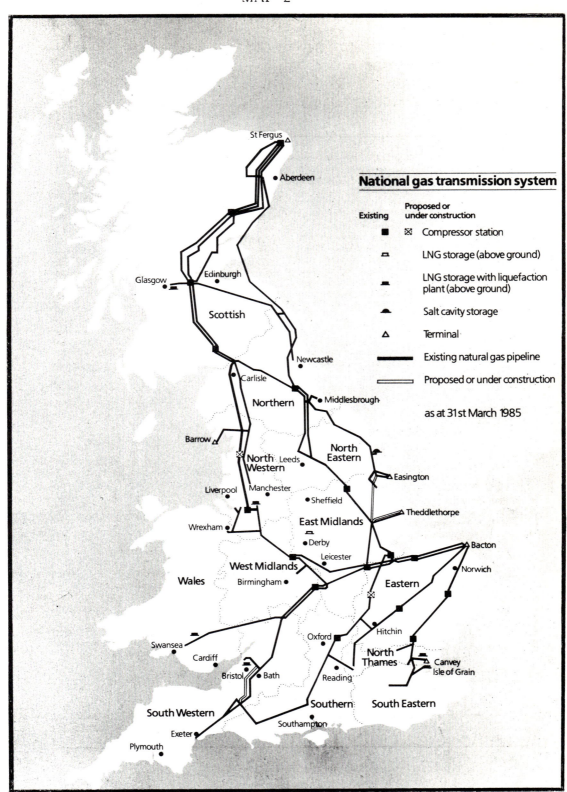

National Gas Transmission System